FIVE
WAR
ZONES

Pergamon Titles of Related Interest

Eliot & Pfaltzgraff RED ARMY ON PAKISTAN'S BORDER
Hunt & Blair LEADERSHIP ON THE FUTURE BATTLEFIELD:
The Complexities of Combat
Joes FROM THE BARREL OF A GUN: Armies and Revolutions
RUSI-Brassey's DEFENCE YEARBOOK 1986
Whelan & Dixon THE SOVIET UNION IN THE THIRD WORLD:
THREAT TO WORLD PEACE?

Related Journals
(Free specimen copies available upon request)

DEFENSE ANALYSIS
DEFENCE ATTACHE

FIVE WAR ZONES

The Views of Local Military Leaders

Abdulaziz bin Khalid Alsudairy
Colonel, Royal Saudi Air Force
Yehuda Bar
Brigadier General, Israeli Army
Lee Suk Bok
Colonel, Republic of Korea Army
Ahmed M. Abdel-Halim
Brigadier General, Republic of Egypt Army
Zia Ullah Khan
Brigadier, Pakistan Army

PERGAMON-BRASSEY'S
International Defense Publishers

Washington New York London Oxford
Beijing Frankfurt São Paulo Sydney Tokyo Toronto

Pergamon Press Offices:

U.S.A.
(Editorial)

Pergamon-Brassey's International Defense Publishers,
1340 Old Chain Bridge Road, McLean, Virginia 22101

(Orders & Inquiries)

Pergamon Press, Maxwell House, Fairview Park,
Elmsford, New York 10523, U.S.A.

U.K.
(Editorial)

Brassey's Defence Publishers,
24 Gray's Inn Road, London WC1X 8HR

(Orders & Enquiries)

Brassey's Defence Publishers,
Headington Hill Hall, Oxford OX3 0BW, England

**PEOPLE'S REPUBLIC
OF CHINA**

Pergamon Press, Qianmen Hotel, Beijing,
People's Republic of China

**FEDERAL REPUBLIC
OF GERMANY**

Pergamon Press, Hammerweg 6,
D-6242 Kronberg, Federal Republic of Germany

BRAZIL

Pergamon Editora, Rua Eça de Queiros, 346,
CEP 04011, São Paulo, Brazil

AUSTRALIA

Pergamon Press (Aust.) Pty., P.O. Box 544,
Potts Point, NSW 2011, Australia

JAPAN

Pergamon Press, 8th Floor, Matsuoka Central Building,
1-7-1 Nishishinjuku, Shinjuku-ku, Tokyo 160, Japan

CANADA

Pergamon Press Canada, Suite 104, 150 Consumers Road,
Willowdale, Ontario M2J 1P9, Canada

First printing 1986

Library of Congress Cataloging in Publication Data

Five war zones.

    Includes index.
    1. Near East--Defenses--Congresses. 2. Near East--
Strategic aspects--Congresses. 3. Korea (South)--
Military relations--United States--Congresses.
4. United States--Military relations--Korea (South)--
Congresses. I. Alsudairy, Abdulaziz bin Khalid.
UA830.F58   1986   355'.033'0048   86-25258
ISBN 0-08-034698-7

Printed in the United States of America

CONTENTS

PUBLISHER'S NOTE

U.S. policymakers, defense strategists, and the American public are often criticized for approaching international problems from the limited vantage point of U.S. experience. They are accused, sometimes justly, of failing to take into account the perspectives of foreign experts who are more intimately acquainted with regional issues and power realities. The long-term result can be a failure for U.S. policies and strategies, and pressures from an uninformed public.

By publishing *Five War Zones: The Views of Local Military Leaders,* Pergamon-Brassey's will extend the audience recently reached by the National Defense University Press of Washington, D.C. Our goal is to provide five high-level foreign officers—a Saudi, an Israeli, a South Korean, an Egyptian, and a Pakistani—with a broader forum for their fresh ideas, thus enhancing the world's knowledge of these critical war zones. This is a unique book: one like it has not been published before. We are pleased to contribute by ensuring the widest possible distribution of these important ideas through Brassey's and Pergamon's worldwide system.

Pergamon-Brassey's
International Defense
Publishers, Inc.
Washington, D.C.
November 1986

A CONFLICT IN THE ARABIAN GULF: AN OVERVIEW OF THE IRAN-IRAQ WAR

Brigadier General
Ahmed M. Abdel-Halim
The Republic of Egypt
Army

CONTENTS

MAPS

INTRODUCTION

The Arabian Gulf region contains half of the world's proven oil reserves. The disruption of this supply in 1973 and again in 1979 profoundly affected world economy. During the past six months, the expansion of the Iran-Iraq War into the Arabian Gulf itself has again threatened interruption of the oil supply.

The stakes in the Arabian Gulf remain high for the United States and its Western allies. US crude oil imports from the Gulf have dropped markedly during the past few years, but there has been no corresponding drop in US interest in the region. Because the industrialized world still depends heavily on Arabian Gulf oil, changes in price and availability could again affect world economy.

The entire Middle East has been deeply affected by the war between Iran and Iraq, the two most powerful Gulf states. The problems that result from the war are not confined to the belligerents but affect all the Middle East, the Arab world in particular, and have had serious repercussions on the international level.

This research paper does not necessarily represent the views of any Egyptian official, the Egyptian Ministry of Defense, or any other Egyptian Department. It only represents the views of the author.

I

THE ARABIAN GULF REGION

The Gulf's Strategic Importance

The Arabian Gulf area has become one of the most important strategic areas on the map of international conflict between the two superpowers. This interest is demonstrated by superpower concern with Arabian Gulf security and superpower attempts to gain the friendship of regional states in order to obtain bases and facilities which would help quick military intervention should the situation warrant it. With both the United States and the Soviet Union intent on securing power in the area, competition to augment their armed forces around and inside the region has been intense. The Soviet Union and the United States are both eager to control the land and naval routes of approach and to better their geostrategic positions.

The Arabian Gulf is an area whose strategic importance has increased with the accelerated flow of oil since World War II. To the Soviet Union, the Indian Ocean represents the historically desired access to warm waters, while the United States considers the Soviet presence in the area a threat to the vital sea lines of communication which bring the allies oil and raw materials.

Technological advancement depends on the availability of oil supplies in sufficient quantity at affordable prices. The Arabian Gulf contains more than 60 percent of the international reserves of excellent quality oil which can be drilled for a relatively cheap price. It is unlikely that economically viable production of new alternative fuels will be achieved in the foreseeable future.

The new, oil-related flow of wealth has created tremendous monetary surpluses, beyond the ability of the oil-producing states to absorb. Consequently, the Gulf has become a centripetal area for international competition to absorb these surpluses. European and American banks have become loaded with Arabian oil money (the "petrodollar"), and the financial stability of nations could be threatened by a decision to nationalize these huge amounts of money with the purpose of either reinvesting them for economic development within the Arab world or linking these credits with political causes.

The Arabian Gulf's geopolitical importance is increased by its proximity to the glacis encircling the Soviet Union. Any US retreat in the area is considered a Soviet advance, and vice versa; therefore, both the US and the Soviet strategies in the Arabian Gulf region are part of their individual global strategies, which take account of the adjacent sea area's importance to international trade.

The United States considers this area within the sphere of influence inherited from Britain in the fifties, and it has sought to reinforce its relationships with most of the ruling regimes. This vital area is considered the protective eastern wing for both the African continent and the eastern Mediterranean. It is also the eastern general defensive line to Europe and the Arab world, and the southern security belt for the Soviet Union.

The low population density of the Arabian Gulf states, and the lack of technocrats with expertise in the different spheres needed for the area's economic and social development have made the area a target for immigration from adjacent states and the Third World generally. Among the foreign communities that have grown up within the region, Iran's is one of the largest—a factor with demographic and other implications. From the military point of view, the Gulf states are weak. Populations are small and scientific and technical cadres capable of handling complicated modern weapons are wanting. This prevents them from developing strong, unified armed forces, despite their acknowledged vulnerabilities to internal and external threats.

Before the start of the current Gulf war, Iraq and Iran were considered the two foremost military powers in the region. Both possessed relatively strong armed forces as a result of their population strengths, and a surplus of petrodollars. Historically, there has long been conflict between Arabs and Persians. Today, mutual antipathy is further fueled by competition to dominate the Gulf area and by intractable border disputes.

The Arabian Gulf region is part of a wider geostrategic area which encompasses the Indian Ocean region and its rim states, and also the Asian states adjacent to the Soviet Union such as Afghanistan, Pakistan, and Turkey. The geostrategic picture is further complicated by the sea lines of communication crossing the region, and the existence of the military facilities of both superpowers within the region.

Energy Assessment

The noncommunist world's petroleum demand declined in 1983 for the fourth consecutive year from an average of 52.5 million barrels a day (mbd) in 1979 to 45 mbd in 1983. Crude oil prices dropped 15 percent in March 1983, from $34 a barrel to $29 for benchmark crude. The decline can be attributed to conservation, a depressed world economy, inventory drawdowns and resulting lower imports in the major consuming countries. Additional oil production, particularly by countries that are not members of OPEC, was another factor. One result of reduced global demand is excess oil production capacity worldwide. In the short run, the single most significant factor which could have a disruptive effect on a surplus world oil market would be a major supply disruption in the Arabian Gulf. At present roughly 7 to 8 mbd of oil transits the Strait of Hormuz from the Arabian Gulf ports. From 1 to 2 mbd additional production leaves the region through the Iraq-Turkey pipeline and the Saudi cross-country pipeline to the Red Sea port of Yanbo. The world depends on from 8 to 10 mbd of oil coming from the Arabian Gulf region. A very small amount of the excess world capacity is located outside the Gulf region.

More than twenty merchant vessels have been hit in the recent Gulf tanker war. At present, this war is having little effect on Arabian Gulf exports through the Strait of Hormuz and the Gulf states have accommodated themselves to the inconveniences caused by the war.

The pressure among Gulf producers is for more production, not less. Iraq is seeking to add to its export capacity by constructing pipelines through Jordan and/or Saudi Arabia and has already completed an extension of its line to Turkey. The United Arab Emirates (U.A.E.), for domestic political reasons, seek greater production. Iran, which has had to draw down its financial reserves, has also sought higher production. Saudi Arabia has faced a precipitous decline in revenues and has had to draw heavily on its reserves to accommodate the production needs of those countries requiring high levels of imports.

An Iran-Iraq peace would bring additional pressure from both countries for higher production levels to finance post-war reconstruction. Opinion varies as to how quickly these countries could increase production levels, but the consensus seems to be that Iran could add to its current production relatively quickly. Iraq, however, would require from 6–9 months to repair its southern export terminal.

In assessing the possible effects on energy markets of an interruption of tanker traffic from Arabian Gulf ports, it is well to keep two factors in mind. First, whether the interruption is gradual or sudden—the longer the market has to accommodate to changed circumstances, the greater the possibility that the disruption can be managed. Second, there are always tankers en route from the Arabian Gulf to American, European, and other ports. Closure of the Gulf does not translate into immediate shortages.

If exports from the upper Gulf (Iraq, Kuwait, Iran) were shut off, this reduction could be accommodated by expanded Saudi and UAE production, and/or additional potential production outside the Gulf. Clearly, a critical actor in such a scenario would be Saudi Arabia. The Saudis have sought, in their recent oil policy, to maintain price stability and they could be expected to be responsive to expanded global demand. Should Gulf exports be completely cut off, from 8 to 10 mbd less would be available to the world oil market. The shortfall would have to be met by the excess capacity available outside the Arabian Gulf, Saudi Arabia's floating storage in the Red Sea, and a joint international effort through the International Energy Agency.

A number of factors would be critical in determining the economic impact of a closure of the Arabian Gulf, including the market's perception of the projected duration of the disruption, the speed with which the additional capacity and reserves could be brought on stream, and the ability of consuming nations to institute conservation measures.

The Superpowers' Interests in the Arabian Gulf Region

British interest during the decades of colonial rule resulted from the Gulf's intermediate position on important routes between the West and India and the Far East. The United States entered the area only after the Second World War. The Gulf's tremendous oil resources have become the focus of US attention as energy demands have increased in Western countries and the region's strategic priority is now second only to Europe in US eyes. The probable US objectives in the Arabian Gulf area are to secure and protect adequate oil supplies from the Arabian Gulf area to the United States and Western allies; to secure the sea lines of communication from the oil exporting ports in the Gulf to the discharge points in the West; to confront the spread of Soviet influence toward the Arabian Gulf; to maintain a special relationship with the governing regimes of the Gulf states; and to establish and maintain a toehold in the forward area that would enable the US to operate against the Soviet Union—hence the deterrent of CENTCOM.

Against the background of the Soviet invasion of Afghanistan the United States declared that "any attempt by any outsider force to gain control of the Gulf region will be repelled by any means, including military force." With the Iran-Iraq war stalemated and trouble in Afghanistan, the tension rose. The United States declared that it would not allow the Strait of Hormuz to be closed. US strategy in the Arabian Gulf area depends on political and military variables and the extent of the probable threat. It is supported by three pillars: the naval strike force; the rapid deployment force; and a forward strategic line consisting of a group of military bases and facilities and strategic cooperation with friendly Arab states.

The Soviet Union is seeking new opportunities in the Arabian Gulf area and throughout the Middle East. Having emerged from their political setbacks there, the Soviets are now in a position to challenge US influence in the region.

Soviet willingness to invest heavily in arms sales and in direct military involvement, together with the weakness attributed to the United States when it withdrew from the Lebanon, have contributed to the improvement in the Soviet position.

The traditional Soviet strategy has been motivated by the desire to reach warm waters and overcome their geographic isolation and to exert influence in the Middle East. Today, the USSR wants to have relations with all the states between the shores of the Indian Ocean and the Soviet border.

Probable Soviet strategic objectives in the Gulf area are to secure in the near future a share of the region's oil to meet increasing Soviet industrial demands; to achieve a balance between Soviet and Western influence in the area; to confront the increasing US military presence in the Gulf and Indian Ocean; to secure its southern borders and support the social, political, and ideological changes within the Gulf states to its own advantage; to control the strategic naval routes in the area; and to establish friendly relations with the Gulf states in order to limit US influence in the region. Estimates indicate that the Soviet Union will, in the very near future, cease to be oil self-sufficient. Soviet planners are emphasizing the goal of procuring needed energy supplies from the nearest source to reduce transportation costs. Clearly, the Arabian Gulf area is the ideal source.

The Soviet strategy in the Arabian Gulf area is subtle enough to adapt itself to local circumstances and variables. The Soviet strategists plan for the long-term and they work in complete secrecy to achieve strategic surprise once a decision has been made to interfere politically or militarily. Within the Gulf area the Soviets depend on: the striking land forces, the Soviet special missions force, the naval force, and the military bases and facilities in vital strategic positions that control the naval and international routes in the area. The Soviets have obtained access to these bases and facilities as a result of strategic cooperation agreements and bilateral agreements with some states of the Arabian Gulf region.

The superpowers remain in competition in the region. Both deploy ships to the Indian Ocean and maintain a military presence in the area. The Soviets signed an arms sales agreement with Kuwait in response to the US arms sales agreement with Saudi Arabia, thus expanding their influence into the Gulf states. There will likely be further Soviet efforts to establish relations with the conservative Arab states of the Gulf. For the moment, the United States and the Soviet Union have some goals in common in the Gulf. Both would prefer to see a negotiated settlement to the Iraq-Iran conflict. Neither has a substantial military force inside the Arabian Gulf area itself. As a result, there is no serious concern that the current Iraq-Iran war could engage the superpowers in hostilities. However, fundamental United States-Soviet interests remain at odds in the Gulf.

The Historical Roots of the Conflict

When the Iraqi armed forces started their invasion of Iran in late September 1980, Iraqi justification centered mainly on regaining the regional rights stolen from Iraq and from the Arab nations as a whole. With this justification, the Iraqi government wrote a new chapter in the bitter historical confrontation, with a legitimate continuation of modern Iraq's several attempts to regain its territory.

The historic roots of today's problem go back to the era of the Persian and Ottoman Empires—and the bitter conflict between Aryan Persians and Semitic Arabs when the border was not clearly defined and changes in tribal affiliations were occurring. A definitive solution

Iran
Administrative
Divisions

—·— Province boundary

⊙ Province capital

0   50 100 150 Kilometers
0   50   100   150 Miles

was never found because of the inaccuracy of the agreements signed between the two empires. A peace and border agreement was signed between the two nations in 1639, but although this agreement gave the Ottomans sovereignty over some territories, the Persians continued in occupation.

The simmering border disputes exploded in 1818 and were centered around Shatt al-Arab. An agreement was signed in 1823, but the likelihood of a new war between the Persians and the Ottomans increased. Britain and Russia intervened and a second agreement was signed in 1837. According to this the Ottomans, under the pressure of the two powers, gave the Persians the city of El-Mahmara (Khorramshahr) in addition to Abadan and the Iraqi territories on the east bank of the Shatt (the Arabestan Region). In return, Persia conceded some territory in the Salaimaneiah province in northern Iraq. Because the two states did not agree on the interpretation of the previous agreements, the Tehran agreement was signed in 1911 and the Constantinople protocol in 1913. The Ottoman government conceded part of al Shatt waters off of El-Mahmara (Khorramshahr). It was determined that the border in this area should follow the middle of the river channel. Thus the principle of freedom of navigation in Shatt al-Arab, and continuation of the Ottoman sovereignty over the waterway was assured.

The recent dispute had several causes. After the emergence of modern Iraq in 1921, the Iraqi claims to Shatt al-Arab were renewed and a complaint made to the League of Nations. The League found no solution and left the whole problem to the concerned parties to settle. In 1937, with the clouds of the Second World War gathering, a settlement between Iraq and Iran (ancient Persia has been called "Iran" since 1935) was a vital matter. The dialogue between the two countries about the borders was resumed and the outcome was an Iraqi concession to Iran. The Thalweg (the middle of the river channel) principle was accepted for 7.75 km from the Shatt before Abadan. Military ships were given a permit to go to the Iranian ports through the river from the Shatt entrance, whereas the rest of the Shatt (less 7.225 km off Khorramshahr and 7.75 km off Abadan) remained Iraqi territorial waters. Three border agreements and an agreement of friendship and dispute settlement were signed in 1937.

The two countries became allies within the Baghdad Pact in 1955, but in 1958 their relationship deteriorated after the Iraqi revolution. In November 1959, the Shah declared that the 1937 agreement was unacceptable, and this impelled Iraq to request sovereignty over the region of Arabestan. In 1969, Iran declared a one-sided abrogation of the 1937 agreement, claiming that this agreement had been signed during the British colonial period and was invalid. Iran now requested that the middle of the river channel be considered as the borderline between the two countries in Shatt al-Arab. The common border in the Khurminshah region where Iran occupies three Iraqi villages on the border between the two countries (Zein-al-Kass, Shukra, and Bir-Ali) was another source of dispute.

In November 1971, Iran occupied the three Gulf Islands which lie at the entrance of the Strait of Hormuz. It also signed an agreement with Oman giving Iran control over the entrance to the Strait of Hormuz in return for sending Iranian military forces to fight against the revolutionaries in the Zaffar region of Oman. On 3 December 1971, Iraq broke its diplomatic relations with Iran, and the beginning of 1972 witnessed several minor clashes of arms between the two states.

The clashes escalated along the borders while the Kurds' activities increased in Northern Iraq. The problem was reviewed by the UN Security Council, Arab and international efforts were made to mediate the situation, and Algiers played an active role which resulted in the signing of the Algerian Agreement on 6 March 1975 which determined that the border line between the two countries is the Thalweg line (the middle of the river channel). On land, the border line was determined according to previous agreements before the British period, with the return to Iraq of the disputed three border villages in the Khurminshah region (these were actually never returned). Both countries were to agree to stop interfering in the internal affairs of the other.

When the Shah, in 1978, asked Baghdad to expel the Iranian dissident, the Ayatollah Khomeini, Iraq agreed, not wanting to violate the Algiers agreement or to undermine stable relations with the Shah. Khomeini's expulsion, after fifteen years in exile in Iraq, seems to be the basis of his enmity toward Iraq and its leaders. As the balance of power shifted in favor of US-armed and supported Iran,

tension between Iran and Iraq increased. Each side supported dissident elements in the other country, and minor border clashes between
the two armies were not uncommon.

After the Shah was deposed, early in 1979, tension between the
two states again increased. Khomeini called on the Iraqi Shia to overthrow the Iraqi government and supported dissident elements conducting terrorist attacks throughout Iraq. Khomeini left little doubt
that he would not be satisfied until an Islamic Republic was established in Iraq under the rule of the Shia. The Iraqis, for their part, did
not welcome the Islamic revolution in Iran. Following the revolution,
the Iranian government faced insurrections in Azerbaijan, Iranian
Kurdistan, Baluchistan, and elsewhere. In these disturbances,
Baghdad took the opportunity to support the dissidents, among them
the Arabs in the province of Arabestan (Kuzistan). Baghdad's fear
was that once the Ayatollah suppressed domestic opposition, Iraq
would be directly threatened. These fears were enhanced by a series
of border incidents.

On 24 December 1979, Iraq demanded a revision of the 1975
agreement and formally objected to the Iranian occupation of the
three Gulf islands. On 17 September 1980, the Iraqi president declared the abrogation of the 1975 agreement and the reactivating of
the 1937 agreement concerning the water border and the 1913 agreement concerning the land border. Iran rejected these claims and the
border incidents started once again, gradually escalating until Iran declared the closure of the Strait of Hormuz on 22 September 1980.
This was considered to justify a declaration of war by Iran against the
Iraqi government. The Iraqi Revolutionary Council declared the same
day that orders had been issued to the Iraqi armed forces to invade
Iran and to attack assigned objectives.

II

THE CONFLICT

The Islamic Republic was formally established in February 1979
after the Shah had been forced to leave Iran in January. From the

very beginning of this new era, Iran's relations with Iraq began to fail and the two countries began making mutual territorial counterclaims. The seriousness of border incidents escalated and widespread expulsion of minorities from both countries began. A rhetorical war between Iranian and Iraqi political leaders simultaneously occurred with verbal attacks and border skirmishes escalating until Iraq declared war on Iran on 22 September 1980. Iraq desired the recognition of Iraqi sovereignty over its own territories, river, and sea waters; an end to the Iranian occupation of the disputed islands; and the termination of Iran's interference in the internal affairs of Iraq and other states in the area.

The underlying objectives of the Iran-Iraq war related to the Iraqi regime's desire to limit the Islamic revolution in order to decrease the spread of revolutionary activities that threaten Iraq and other regimes in the Gulf area. In addition to the personal enmity between Khomeini and President Saddam Hussein, strong mutual suspicion existed between the Iranian and Iraqi regimes.

When the Iran-Iraq conflict began, Iran was suffering from internal political troubles, economic deterioration, military weakness, and a worsening of diplomatic relations with several countries in the world. These circumstances offered Iraq an historical opportunity, that might not be repeated, to implement its claims.

The Status of the Belligerents

Iraq sought to implement political preparedness on three levels: internally, in the Arab world, and internationally. Internally, all the centers of organized political opposition were greatly diminished; a reorganization of the Baath party was carried out, and the minorities, especially the Kurds and the Shia leaders, were controlled. Within the Arab world, Iraq reinforced its relations with the Gulf states, in particular with Saudi Arabia. It also instituted political and economic cooperation with Jordan, which is thought by Iraq to provide it with strategic depth. On the international level, Iraq reinforced its relations with the Western countries, especially France, and tried to approach the United States by supporting the US stands against Iran. Iraq hoped to gain the support of the United States or, failing that, to neutralize the Americans.

Iraq prepared economically as well as politically for the war. Over the years, Iraq has used its petroleum resources for a very ambitious economic plan. The existence of the necessary financial infrastructure and the complete control of resources by the Iraqi government helped in the swift implementation of this plan. Schemes were elaborated and efforts were exerted to bypass any material or economic shortages. Huge quantities of goods and food were imported and stockpiled.

Iraq used all possible diplomatic means to make known its claims against Iran. On 24 December 1979, it officially requested that Iran revise the 1975 agreement, and the Iraqi Foreign Ministry sent official messages to the Secretary General of the United Nations clarifying the Iraqi position regarding the illegitimate Iranian occupation of three Arabian islands in the Strait of Hormuz.

Iraq began to consider the possibility of war against Iran at the time of the 1979 Iranian Revolution. Iraq prepared the domestic front for war with Iran by exploiting the circumstances and obstacles facing the Iranian revolution. Steps were taken to bolster morale in Iraq, and in a wider Arab connotation. Internally, the government stimulated Iraqi nationalism by reminding the people of Iraq's historical rights in Shatt al-Arab and the province of Arabestan (Khuzistan), and other territorial areas adjacent to the border occupied by Iran. The government also reminded the Iraqis of several occasions when pressure had been exerted on Iraq by the deposed Shah. It was not difficult to provoke a tremendous upsurge of feelings of outraged dignity.

Iraqi propaganda concentrated on provoking Arab feelings regarding the occupation of the three Arab islands. It also opened the historical file to stimulate Arab enmity towards the Persians. Iraq reminded the Arabs of Iranian support to Israel, especially in the sphere of military cooperation. Iraqi propaganda underscored the Iranian revolution's goal of constructing an Islamic Shia Empire at the expense of the Arab states.

Meanwhile, military preparedness concentrated on command, training, and armament. Military commanders with the required expertise, were given high military commands, regardless of their political affiliations. They were given full authority to exercise command over their units according to a general plan of preparedness. The Iraqi armed forces completed tactical and operational command exercises according to scenarios for the expected war with Iran.

Iraq supplied its armed forces with large quantities of advanced and diversified weapons. The armed forces completed training on those weapons, supported by the Iraqi cooperative relations with various Eastern and Western bloc states. This was done in the framework of a diversified weapons plan supported by the necessary finances provided by the surpluses from Iraqi oil exports.

The Iranian Revolution disturbed the regional balance: the Revolution supported the concept of an Islamic league, instead of the Arab nationalism movement. Despite the threats of the revolutionary commanders to export their revolution to neighboring states, Iran was not ready for war for a number of reasons. The internal situation in Iran was complicated by the existence of different nationalities. Support for the Revolution was reduced in most levels of society as a result of unemployment and the ongoing struggles for political power. The economic situation was deteriorating simultaneously with the political situation after the American hostage crisis.

The formation of the Revolutionary Guard and the effect this had on military discipline resulted in a reduced military capability. The low level of training, the lack of proper equipment maintenance, and the want of spare parts following the cancellation of arms agreements were other factors that weakened the military. The expulsion of most of the officers above the rank of brigadier general, the series of courts martial of various commanders, and the flight of others further depleted the officer ranks.

The Islamic Republic Party, headed by Ayatollah Khomeini, is the predominant political organization in Iran. The party controls an overwhelming majority in parliament and is supported by the Revolutionary Guard (the party's military wing). The civilian wing consists basically of intellectuals and technocrats. The competition within the party between civilian and religious wings ended in the victory of the latter. At the beginning of the war with Iraq, racial minorities were disaffected and disputes over authority precluded unified political decisionmaking.

After Khomeini's declaration that the monarchical system contradicted Islam and that Bahrain was an integral part of Iran, the ruling circles in the Gulf states feared the possibility of revolution being exported to their territories. Iran also broke its diplomatic relations with Egypt, continued its occupation of the three Arab islands, withdrew its military support from Oman, and allowed its relations with the surrounding Arab states to deteriorate. On the international level, Iran lost the sympathy of most countries—not the United States alone—as a result of the hostage crisis.

The Iranian economic situation has deteriorated since the Revolution. The United States has reduced its oil imports while the Soviet Union has reduced its gas imports as a result of the situation in Afghanistan. Oil production has decreased by fifty percent, natural resources have not been exploited economically, and economic relations with the Gulf states have deteriorated. The poor economic situation caused a cancellation of military equipment agreements with the United States amounting to 9.2 billion dollars.

Iranian diplomacy was handicapped in its efforts to improve Iran's image on the international front by the Ayatollah's fanatical diatribes against the United States and the Western world. Threatening the Gulf states, awarding arbitrary verdicts in the revolutionary courts and, above all, taking US diplomats hostage all diminished Iran's standing.

As a result of internal disputes over authority, increased unemployment, executions, and the deterioration in the political and economic situation, civilian and military morale was at a low ebb at the start of the war with Iraq. Before the Revolution, the Shah had depended on the armed forces to control the Iranian people. The Revolution's new leaders accordingly reduced the role of the armed forces, weakening its combat capabilities. This weakening was clearly shown in the low level of training (the result of constant changes in the commands and headquarters), and the lack of technical cadres and trainers. The success rate of those in training fell to approximately 40 percent in the wake of the establishment of the Revolutionary Guard. Military discipline and control over the armed forces declined with the removal of many former officers. In sum, Iran faced the Iraqi offensive ill-prepared politically, economically, diplomatically, morally, and militarily.

The Belligerents' Objectives

The President of Iraq's declaration in February 1980 contained eight political points indicating his country's objectives. Iraq wished to be the dominant state in the Gulf area nationally, politically, and economically. Iraq also wished to take advantage of the hoped-for renewed supremacy to mount ambitious development plans while trying to decrease or eliminate the influence of the superpowers in the region. The political-military objective was to regain the Iraqi territories in Shatt al-Arab (Khuzistan) and other occupied territories while securing the political borders with Syria. The goal of the strategic offensive operation was to defeat the main concentration of the Iranian armed forces on the common borders; to recover Iraqi lands in Shatt al-Arab; to profit financially from the resources in the province of Arabestan; and to overthrow the Iranian regime, or at least to control it sufficiently to secure Iraqi national security.

To achieve the goal of the strategic offensive operation the Iraqi forces were to launch a major blow against the Iranian forces in the area of their common borders in the eastern strategic direction, with the immediate mission of defeating Iran and occupying a line-in-depth. As a final mission, the plan was to develop combat operations to complete the defeat and destruction of the enemy, and occupy the whole province of Arabestan. The army's main efforts would be concentrated in the southern operational direction, while assuming the defense on the borders with Syria. A total defeat of Iran would secure, once and for all, the Iraqi claims.

Imam Ayatollah Khomeini's speeches defined Iranian goals. The goals were the strengthening of Islam, Iranian independence, and the uniting of all Islamic nations. The plan centered on the unity of Moslems and the steps to be followed to reach unity. The Ayatollah's speeches determined the Iranian political goals and objectives and, later on, the operational goals and the missions of the armed forces as well.

Politically, the Ayatollah aimed to spread the Islamic concept, making Islam the spearhead of the fight against Zionism, imperialism, and communism. He wished to decrease or eliminate the influence of the superpowers in the region, and to make Iran a non-aligned country. His political-military objectives were to secure the Iranian revolution internally, to protect the religious oligarchy, to propagate the comprehensive Islamic concept among the regional states and, by using Iranian armed forces, to secure political borders and oil resources.

The goal of the Iranian strategic defensive operation was to halt the Iraqi incursion and to prevent dispersion of Iraqi forces followed by the attrition and defeat of any Iraqi troops who had succeeded in penetrating the borders. Ultimately, Iran wished to create favorable conditions to launch a counterattack to regain the occupied territories as far as the border. A further aim was to threaten Iraqi strategic targets and Iraqi oil resources.

The Phases of the War

The Iran-Iraq war has passed through several phases. The preparatory phase lasted from 11 January to 15 September 1980, and was characterized by intense media campaigns in both Iran and Iraq and by the spread of border incidents and insurgencies.

Initially, the deteriorating situation between the two states was reflected in the insurgencies on the borders and the launching of artillery fire. The situation entered a new phase after April 1980 when border incidents and terrorist attacks became an almost daily event in different places both along the borders and in depth against military and civilian targets. From the beginning of September 1980 Iraq exploited the escalating situation. It deployed its air force in the daily struggle, and readied its forces on a different axis for strategic deployment in a later offensive. At the start of the fighting, Iraq had a definite quantitative superiority—and a degree of qualitative superiority—over Iran. It goes without saying that the Iraqi command's estimation of the Iranian side led it to suppose that the war would be one of lightning strikes; it was anticipated that Iraq would be able to implement its goals and withdraw its armed forces in a limited time. The events of the war proved otherwise.

The second phase of actual fighting began when Iraq used armed force to compel the Iranian political command to recognize Iraq's legitimate rights. The main strategic operation, from mid-September to mid-November 1980 was followed by the attrition phase lasting from 17 November 1980 to 26 September 1981. The period from 27 September 1981 to 17 March 1982 saw Iran regaining the initiative and the start of counter-operations. From 18 March to 12 July 1982 territory was regained by Iranian counterattacks and, from 13 July 1982 to the present, there have been continuous Iranian efforts to mount a general offensive against Iraq.

The Iranians estimated that Iraq would not launch a comprehensive offensive against Iranian territory. As a result, Iran neglected to prepare the nation, the armed forces, and the operational theater for war with Iraq, thinking the fighting would not escalate beyond border incidents.

In the opening phase of the actual fighting, the Iraqi forces began the main offensive operation on 16 September 1980 in the direction of the disputed borders, occupying Zein-Alkas, Al-Shurkjra, and Bir-Ali in the area of Kasr-Sharin in the northern sector. On 22 September, Iran declared the Iranian land, naval, and air borders with Iraq a combat area, in addition to stopping navigation in the Strait of Hormuz. Iraq considered these acts as an Iranian declaration of war.

At 1100 hours that day, the Iraqi air force directed a concentrated air attack against economic installations, populated areas, and military targets in Iranian cities. Under cover of the air attack, the Iraqi land forces penetrated the Iranian borders, directing a main blow in the direction of the southern sector and two secondary blows in the direction of the intermediate sector. In the northern sector they developed an offensive and occupied some other cities. The cities of Khorramshahr and Abadan were encircled on September 19, Kharramshahr was occupied on October 24th. The Iraqi armed forces were unable to occupy Abadan because the Iranian command quickly reinforced the port's defenses. This was one of the main factors that changed the course of the war to the benefit of Iran. During this time, the Iraqi forces succeeded in dominating a strip 800 km long and 20–60 km in depth, that extended from Khorramshahr in the south to Kasr Sherin in the north.

During the attrition phase, combat operations were characterized by air and artillery strikes on both sides, by the success of Iraqi forces in holding occupied terrain, and by the failure of all the counterattacks by Iranian forces. As the war entered its next phase, the Iraqi force did not attempt to modify its positions. The city of Abadan was encircled, but no attempt was made to enter it.

Iran exploited this lull with a concentrated effort to reorganize its armed forces and to move internally and externally into a state of war. The forces along the front were repositioned, and the necessary maneuvers to the different sectors were carried out. Release of many military leaders from jail allowed the army to exploit their expertise in military operations. Volunteers were allowed to join the Revolutionary Guard and reinforcing, equipping, and training of these forces to prepare them for action side by side with the army went ahead. Initially, some limited offensive operations by the land forces with parallel air and artillery strikes against the Iraqi forces were undertaken, while widespread preparations for operations to regain the occupied territories were underway.

In the period from 27 September 1981 to 17 March 1982, Iran regained the initiative, beginning counteroperations. On 27 September, the Iranian forces launched a counteroffensive in the southern sector. They raised the siege of Abadan and forced the Iraqi forces to withdraw towards Khorramshahr. The Iranian forces also launched other counterattacks, liberating most of the occupied cities. Iraq kept its main forces on the front line without either suitable defenses in depth or reserves. This led to the success of the Iranian tactic of using masses of people within limited sectors and penetrating to the area line to regain considerable territory. Iraq withdrew the main concentration to the rear, concentrating the defense on vital lines. A great number of the Islamic Revolutionary Forces (approximately 60,000) were committed to fight alongside the Iranian forces. The Iranian forces applied many initiative-concentration-surprise-maneuver tactics with good effect. Their frequent night operations contributed to their success. A succession of Iranian counterattacks began in March 1982, forcing the Iraqis to retreat from much of the territory they had occupied. The Iranian success underlined the revitalized spirit of the Iranian armed forces, Iran's greater resources of manpower, and better coordination between the regular army and the Revolutionary Guard.

On 18 March, the Iranians launched a secondary counterattack against the Iraqi forces in the intermediate sector, with the aim of deceiving the Iraqis as to the direction of the main blow.[2] This operation was called "Fatma-al-Zahraa."*

On 22 March, the Iranian armed forces started the main blow of the operation "Fateh" in the intermediate sector.[3] The Iranian forces

---

* See Endnotes for the significance of these terms.

were able to regain roughly 20 km to the west on a front of 100 km. Eight days after the renewed fighting began, as a result of heavy Iraqi casualties, President Hussein issued his orders to the Fourth Corps to modify its positions by retreating to suitable lines inside Iraqi territory.

Three counteroffensives were launched in the course of the first stage of the "Bit-al-Makdes" operation.[4] A main blow was launched in the direction of Khorramshahr and a secondary blow, in the direction of the Karon river, attempted the crossing of the Karon river and an advance in the direction of Khorramshahr to establish contact with the forces advancing from the Ahwaz direction. Another secondary blow, with the goal of striking the advance Iraqi reserves, was undertaken.

The Iraqi forces were able, in the first stages of fighting, to contain the Iranian attack. The Iraqis were also able to halt the advancing forces from Ahwaz to Khorramshahr with high density air and artillery attacks, causing heavy casualties to the attacking Iranian forces. After eight days Iran, determined to succeed, committed fresh forces through the gaps that had been made in night operations, and continued its pressure on the defending Iraqi forces. On 8 May, the Iraqi command issued orders to withdraw the forces to new positions (a regrouping operation) to create the most favorable conditions for these forces to launch a successful counterattack against the Iranian forces in this sector reinforcing the defenses of the port of Khorramshahr. The Iraqi decision to withdraw its forces to new lines caused a relative lull in combat operations in the southern sector and a decrease in the combat actions ratio, especially after the Iranian forces' partial success.

In the second stage of the operation, Iran regained the city of Khorramshahr. The Iranian command declared that it was essential to liberate the occupied territories. In response, the Iranian forces continued their southward pressure on the Iraqi forces and were able to regain the province of Arabestan (Khuzistan). On 24 May 1982, the Iranian forces regained the previous position on the borders. At the end of the day, Iraq was still occupying a strip of land 710 km long and 20–40 km wide along the borders. After these operations, the situation was characterized by relative quiet though air and artillery strikes by both sides against military and civilian targets continued.

On 10 June 1982, Iraq announced its willingness to end the hostilities between the two Islamic states. Moreover, the Iraqis stated their intention to withdraw within two weeks from all occupied Iranian territory and declared their willingness to accept binding arbitration by the Islamic Conference. Iran rejected the offer as "too late," and not providing for the removal of President Saddam Hussein. Nevertheless, the Iraqis withdrew their forces from Iran as a sign of their good intentions and to gain moral high ground.

In July 1982, the Iranians renewed their attack on Iraqi territory. For two years, Iraqi forces had been primarily defending their own territory while inflicting heavy losses on attacking Iranians. Some very minor advantages had been gained by Iran, but at a tremendous cost in lives and equipment. Both of the belligerents demonstrated their willingness to pay the high cost of victory and to use all necessary means to attain it.

The Present Military Situation

The military balance has shifted in Iraq's favor during the past two years as a result of the worldwide arms embargo on Iran, and massive French and Soviet arms sales to Iraq. Iran has received virtually no major military items in two years and has had to cannibalize spare parts to keep equipment operating. The Iraqis, on the other hand, have in the past years received T–72 tanks and advanced MIGs from the Soviets and Super Etendard aircraft with Exocet missiles from France. As a result, Iraq enjoys a significant advantage in operational fighter aircraft, armored personnel carriers, and tanks.

Since withdrawing from Iranian territory in 1982, Iraqi forces have successfully withstood repeated Iranian offensives. Iranian human wave assaults have been ineffective in the face of superior Iraqi firepower and tactical air operations. On the southern sector, where any new Iranian offensive is expected to take place, the Iraqis have made good use of delays by fortifying defensive positions, laying minefields, flooding possible attack routes, and replacing weapons lost in previous battles.

Iraq does have a potentially serious military liability. This is its lack of strategic depth. Basra, Iraq's second largest city, is literally on the front line. While most observers believe Iraq can prevent an

Iranian breakthrough, the possibility of a breakthrough cannot be totally discounted. Should Iran's armies penetrate Iraq's defenses, they would only have to go a few miles to sever the Baghdad-Basra road and to threaten Basra itself. To win a major victory, Iran would have to rely on its ground forces to exploit Iraq's liability. Most ground assaults thus far have come from the Revolutionary Guard. A coordinated attack with the army might be able to penetrate some weak spot in the Iraqi lines. Were such a breakthrough to occur, it could conceivably affect Iraqi morale and decisionmaking.

It is my assessment that an Iranian assault on Iraq would probably lead to a defeat for Iran. Should Iran fail in its assault, its stocks of material would be almost exhausted and the risk of an Iraqi counteroffensive would exist. Given this assessment, Iran faces a dilemma. Iran has suffered the loss of more than half a million killed and wounded in the war thus far and the lack of victory could destabilize the government. Iran has publicly heightened expectations for a major assault in the next few months, but Iran's military leaders probably share a pessimistic assessment of their chances. The debate continues in Tehran and is complicated by the Ayatollah Khomeini's poor health.

Both Iraq and Iran have a series of military options available to them. Iran's includes an assault on Basra, closing the Gulf to shipping, attacking the Gulf Cooperation Council (GCC) states, and escalating terrorist acts. None of these actions could be taken, however, without placing some Iranian interest in jeopardy. Iraq's major military mission is to end the war quickly, which means withstanding Iranian ground attacks, and reducing Iran's economic advantage by escalating the tanker war and possibly attacking Kharg Island. Because of recent French and Soviet arms sales, Iraq now has the capability to inflict severe damage on Kharg Island's oil terminal. Iraq has probably refrained from the attack because of the military cost, and the political advantage of holding a threat over Iran's head.

The Tanker War

More than twenty merchant vessels have been hit in the recent Gulf tanker war which originated in Iraq's desire to limit Iran's oil revenues and to internationalize the war. While Iraq, using Exocet missiles and Super Etendard aircraft has been successful in focusing world attention on the war, it has not halted the flow of oil from Iran or the rest of the Gulf. Iran has retaliated, making Iraq and its Gulf supporters pay a price for Iraqi attacks in the Gulf.

The Iraqi attacks were targeted against ships bound for or departing Iranian ports, and they took place primarily in the Iraqi-declared Exclusion Zone or in Iranian waters. Iraqi attacks outnumber by two to one the number of Iranian attacks and are generally more damaging. The French-made Exocet missiles often strike in the engine room area whereas the less powerful American-made TAV-guided Mavericks strike the superstructure. However, according to press reports, the Iranians have recently successfully used radar reflecting decoys to confuse Iraqi pilots.

Iranian attacks have generally been in response to Iraqi attacks, but their targets are ships bound for or coming from Kuwaiti and Saudi ports. Iran's F–4s attack during daylight hours with plenty of warning by reconnaissance aircraft. Iran's targets were in the Upper Gulf until Saudi Arabian F–15s destroyed two Iranian F–4s near Jubail. After the Saudi response Iran shifted all its efforts to the Lower Gulf, beyond Saudi air cover.

If the stalemate in the Iran-Iraq war continues without a diplomatic settlement, however, there is some prospect for an escalation in the tanker war. The spectrum of escalation could range from more air attacks on tankers by both sides to an attempt by Iran to close the Gulf altogether. Closing the Gulf would be an act of desperation on Iran's part because it relies heavily on the Gulf for imports and exports, and because after such an event, the GCC states would be less reluctant to take action against Iran and big power intervention would be likely. Unless the major powers intervene to prevent it, Iran does have the capacity to close the Strait. Iran has an adequate number of World War II vintage contact mines which would probably be placed in the shipping lanes of the Lower Gulf, rather than in the Strait itself. In addition, Iran could use its aircraft, its land-based artillery, and hundreds of small explosive-laden speed boats to intimidate traffic.

III

THE IMPACT OF THE IRAN-IRAQ WAR

Local Results

The Iran-Iraq war has had political, economic, and military results that were not anticipated by either belligerent. The Iraqi regime estimated that a lightning campaign, and a swift defeat of the Iranian military would lead to insurgency on the part of the several minorities that constitute Iran, but this did not happen. Iraq's error lay in not realizing that the appearance of an external threat would inspire unity and a will to carry arms to defend Iran. The same result occurred on the opposite side. Arab minorities in the province of Arabestan (Khuzistan) did not support or welcome the Iraqi forces as Iraq had expected.

Iran's miscalculation lay in picturing the Iraqi regime as weak internally and politically unpopular. Iran, aware of the many strong opposition movements, imagined that the Shia would immediately respond to the appeal of the Iman Khomeini, but this did not happen either. National affiliation apparently outweighed, in time of danger, minority affiliations. In any event, all Iraqi political factions seemed united in the upsurge of nationalist feeling stimulated by the war against Iran. Such major miscalculations by both regimes led ultimately to the destruction of the economic centers of both countries. Today, the unyielding attitudes of both belligerents prolong the war while the situation deteriorates and neither side can achieve complete superiority. Attempts to hamper navigation in the Strait of Hormuz and the Arabian Gulf may lead to the direct intervention of the two superpowers and other big powers in the region.

The Iran-Iraq war has become a war of attrition for both countries that now have diminishing resources and weakened economic positions. This lengthy war has led, in turn, to loss of control over the Gulf states while their economic support remains necessary to the two belligerents. It was the accepted wisdom that the world could not do without every drop of Gulf oil and would never survive the desolation of any of the oil producing countries. The contrary has been proven. Despite the disruption of two oil producing states (representing one-eighth of world production) Western economies were not affected. The two belligerents were the ones who suffered. As a result of the disruption in Iraqi and Iranian oil supplies, the rest of the oil producing countries in the Gulf region increased their oil production capacity. This increase has led to an oil surplus and price reductions. The oil producing countries, meanwhile, failed to regain the total unanimity they had had in 1973. Instead of oil constituting a weapon in the hands of the two belligerents, it turned out to be a dagger in their backs, depriving them of considerable revenues. The war between two big OPEC states eventually weakened the oil producing and exporting states, not the operations of international oil manipulation.

At the outset, the Iranian revolution crushed Israeli influence in Iran and sought to confront Israel directly by giving unlimited support to the Palestine Liberation Organization. As a result of the war Iran was obliged later to deal with Israel and Syria simultaneously. This led to the unexpected alignment of Israel and Syria, for different reasons, in one front against Iraq. One of the important military-political results of the war was the change in Arab support. Iraq, having been one of the sources of Arab support, became one of its recipients. Gradually, the Gulf States' interest shifted, temporarily, from confronting Israel to the immediate concern with the Gulf's regional equilibrium.

The Iran-Iraq war could be prolonged for years unless a continuation of the struggle conflicts with the superpowers' interest. However, the war may be coming to an end because diminished economic strength and fighting capability may prevent either party from acting decisively to end the war. The most important objective of the war has already been achieved: Iraq and Iran are no longer powerful enough to threaten others. There are three possible conclusions to the war. First, Iraq's victory and Iran's defeat; this seems unlikely since Iraq has completely lost the initiative. Second, Iran's victory and Iraq's defeat; Iran's internal circumstances and the solidarity of the Iraqi armed forces make this an improbable solution to the conflict. In addition, the international powers are unlikely to permit the conflict to end with victory for the Islamic revolution. The third alternative, a stalemate, is the most probable ending.

The Iran-Iraq conflict has now passed the stage of military conflict, and escalated to an economic struggle, with the belligerents seeking mutual destruction of economic and oil installations. Iran estimated two weeks after the beginning of the war that the struggle was costing approximately $25 billion more than a year's revenue from oil exports. Iraq estimated the cost in the same period at $12 billion. These estimates relate to the direct costs of war and do not include the opportunity costs resulting from decreasing, and then halting oil exports. The increase in import costs, especially transportation costs, has been steep for both countries. Basra, once Iraq's major port of entry, is now subject to heavy Iranian air and artillery strikes.

It will take some time after the end of the war for either Iran or Iraq to reach their previous oil production levels. Both countries' economies will suffer at least until the year 2000 while they rebuild facilities destroyed by war. The only nations to benefit from this situation will be the international powers, especially those in the Western Hemisphere, who still control modern technology.

The Iraqi economy has suffered from Iranian strikes against oil and industrial installations. In addition, the agricultural sector has been hurt, and Baghdad's electric generating plant has been partially destroyed. Agricultural production in the province of Shatt al-Arab has stopped completely. With a reduced GNP, Iraq basically depends on the Gulf states' support and loans. As oil exports have fallen with the halting of tanker traffic through the port of Basra, Iraq has come to depend on the pipeline through Turkey and has sustained a total annual loss that amounts to $4 billion.

Iraq will be exposed during rebuilding to strong external pressures since its oil revenues at this stage will be insufficient to sustain its needs. Iraq is going to need external financing and technical expertise, but it has lost the international political clout which enabled it to secure several privileges in the late seventies. To rebuild its production capacity to its pre-war level, Iraq may be forced to draw back within itself for a period that may last as long as five years, and this could result in some constraints on the possibility of widening its regional role, especially in trade and industry.

In the military sphere, the war will have important aftereffects in addition to those already noted. As a result of its limited population, Iraq was forced to form a people's army for the local defense of vital Iraqi targets. Increased military cooperation between Iraq and the Western countries, especially France, had a parallel in increased military cooperation with the Soviet Union, which supplied Iraq with T–72 tanks and some other arms and equipment. Iraq will basically depend on the Soviet Union to rebuild its armed forces, unless a political decision is reached to diversify arms sources. Iraq's economic requirements can only be satisfied by the West.

While there will probably be increased political opposition, the Iraqi Baath party's control over all the state's institutions will likely endure. Opposition to President Saddam Hussein exists, but he has the support of the Iraqi people and the armed forces. The continuation of the Gulf states' financial support to Iraq will cushion the impact on the Iraqi people of the economic aftereffects of war. Iraqi policies concerning the Palestinian problem have changed. This is shown by the ending of its strong rejectionist attitude and the new closeness and cooperation with Egypt, despite the absence of diplomatic relations between the two states.

Iraq is showing new understanding toward the unsolved border problems with some neighboring states—Kuwait, Saudi Arabia, Jordan—and has assumed a new moderate policy toward the Arabian peninsular states with the aim of obtaining economic support and the needed finance to support its deals. These developments have weakened Iraq's influence on the Gulf states and strengthened the Saudi role.

On the surface there is little sign in Baghdad today of the epic struggle going on some 100 miles east of the city. Blackouts, which characterized the first year of the conflict, are a thing of the past. A striking new feature of the Baghdad landscape is a series of architectural landmarks completed after the war began. But below the surface there are signs of war weariness. There is some evidence of increased maturity among the Baath regime in Baghdad and in their thinking. They are now more practical and pragmatic; they recognize that the

West can best assist in the modernization of post-war Iraq and that the moderate nations of the region, Egypt, Jordan, and the GCC states have been most helpful to Iraq during its war with Iran.

The Iraqi economy remains in serious but manageable shape. The problems are many: deferred payments offered by creditors; continued subsidies by GCC countries (including the sale of oil for Iraq's account by Kuwait and Saudi Arabia); the need to achieve increased petroleum exports through the Turkish pipeline; and the imperative for new pipeline construction to ensure additional oil exports. Iran's strategy of crippling Iraq economically by cutting off Baghdad's ability to export petroleum through the Gulf has not been fully successful, but Iran has succeeded in derailing Iraq's massive effort to modernize the industrial infrastructure before 1983. The critical economic issue facing Baghdad at present is whether it can expand its petroleum exports despite the closure of the Gulf to its tankers and the refusal of Syria to permit exports through the Syrian pipeline. Three possibilities exist: the Turkish line has expanded production from 700,000 barrels per day to 1 million barrels per day and another line is under discussion. The Export-Import Bank has agreed to provide substantial guarantees of financing for a pipeline running through Jordan to the Gulf of Aqaba, and the possibility of constructing an Iraqi spur to the east-west, cross country Saudi pipeline is being considered.

Iraq's biggest supporters in the war are the moderate Middle East states like Egypt, Jordan, Saudi Arabia, and Kuwait. In contrast, Iraq's fellow-rejectionists, Syria and Libya, have supported Iran. Iraq's new-found alignment with the moderate Arab states may have, in the long term, a moderating influence on Baghdad. The US position on the war has led to the resumption of diplomatic relations between the two countries. Moreover, Iraq has modified its position in the Arab-Israeli conflict from the refusal to consider any peace initiatives to a stated willingness to accept any settlement agreed to by the Palestine Liberation Organization (PLO).

Iran will need the same period for rebuilding the basic economic infrastructure destroyed by war. Continued low productivity resulting from a want of raw materials will likely lead to increased inflation, raise the cost of living and unemployment figures still further, and will prolong Iran's economic deterioration which began with the revolution. Iran's economic rebuilding will require modern technology which is found in the United States and other Western industrial states. This requirement may lead to a new, moderate Iranian attitude in contrast to the situation which reached its peak during the hostage crisis. Iranian economic deterioration has caused the unemployment ratio to reach about 40 percent of the labor force. Other results have included price increases of 300 percent, a diminution of imports and exports, and a reduction in Iranian oil production from 2.5 million barrels per day to 400,000 barrels per day. The destruction of the Iranian economy has curtailed the leadership's ambition to continue as the most powerful state in the area. Iran's external savings have been seriously depleted, and political and economical freedom was lost to Iran when it was compelled to seek external loans.

Heavy casualties in arms and equipment, especially in tanks, have impaired Iranian military capabilities. The loss of most of its technical cadres, which alone were adequately trained to maintain the technical efficiency of arms and equipment, will seriously undermine Iranian capabilities. As a result of the war, Iran's direct and indirect vulnerability to Soviet invasion has increased. The newly-established Revolutionary Guard gained experience under arms and will form the nucleus of the post-war revolutionary armed forces. The war has resulted in the unification of all Iranian social and political organizations in the face of the Iraqi invasion. In this time of tension the Khomeini regime, besides directing the war effort, has succeeded in controlling Iran and eliminating political opposition.

Needing external political and military support, Iran has made an effort to better its international relations. Gradually, as Iran-Soviet relations started to evolve, partial success has been achieved in securing economic and military supplies from several countries. Between the East and the West, Iran has raised the banner of non-alliance. Its relationships with the United States and the Soviet Union are characterized by wariness. Iran is trying to establish a balanced relationship with the two superpowers. Meanwhile, its relationship with the neighboring Arab states is still characterized by tension.

The Iranian regime is still working to export revolution outside Iran. To achieve this objective, political asylum is offered to revolutionaries coming from abroad, and an effort is made to recruit sympathizers through Iranian embassies overseas. The failure of Iranian attempts to penetrate Iraqi territory has had some negative results within Iran. Differences between the war supporters and pacifists may end the detente among the opposing powers and might rekindle resistance activities.

Several key Iranian leaders appear to be reaching the conclusion that the costs of continuing military efforts are becoming too great. Tehran's hesitancy about launching an offensive at Basra reflects, in part, its concern that any gains on the battlefield may be outweighed by heavy casualties which could, some Iranian leaders fear, spark strong domestic reaction. Dissatisfaction with the war has increased in recent months. Iranian leaders may be worried that this dissatisfaction, coupled with continuing economic hardship, power and water shortages, and general governmental inefficiency, could trigger antigovernment disturbances and demonstrations. As yet, however, no opposition group has emerged capable of fashioning significant antigovernment political activity.

The Iranians are aware that their aircraft are no match for the Saudi F–15s and that an escalation of terrorism could reinforce the Gulf states' fears of the export of Iranian revolution and ensure their support for Iraq. With limited options, an Iranian diplomatic effort seems to be underway to separate Iraq from the Gulf states and to gain some influence in Moscow. Iran wants to reduce the amount of economic and political resources going to the support of Iraq from the Gulf states. Consequently, the Iranians appear to be lessening their anti-GCC actions. Recently, there have been low-level exchanges between officials from Moscow and Tehran. These have, as yet, led nowhere and would first have to overcome Khomeini's objections to dealing with Iran's communist neighbor.

Iraq's efforts to damage the Iranian economy are having some effect. Although the Iranian regime has considerable financial reserves, these have been declining since early 1984. Iraqi attacks have increased the tempo of this decline. Spot shortages are occurring. Although Iran continues to procure some weapons and ammunition from a variety of sources, including North Korea, Libya, Syria, the Eastern bloc nations, and some Western bloc nations, weapons procurement is becoming increasingly difficult and expensive.

Regional Results

The war has had repercussions throughout the Middle East—in the Gulf States, the PLO, Israel, and Egypt. The increased tension and weariness among the Gulf states that result from the war constitutes an invitation to foreign influence in the area, especially that of the superpowers. With the ending of the war, these influences will be reinforced, putting added strains on regional stability. The political and social status of the ruling regimes of the Gulf and Middle East area will also come under stress.

A peaceful settlement of the conflict, after Iraq and Iran's capabilities have been reduced, will liberate the Gulf's rulers from their fears. Were the war to end in Iran's favor, Saudi Arabia would be the next target. The stability of the Saudi regime, and consequently the Arabian peninsula and the Gulf states, would be directly affected. The Iran-Iraq war is one of the most important factors contributing to the cooperation and coordination among the Gulf's states. For the first time comprehensive security assessments have been made, and an indigenous Gulf organization to defend the Gulf (GCC) has been created. At the same time, military relations between Egypt and the United States have been strengthened to defend the region's countries.

At the start of the war, Iraq was supported by most of the Arab countries and, in particular, by the Arabian Gulf states. Both Syria and Libya stood beside Iran, while Algiers played the role of a mediator between Iran and the United States in solving the hostage crisis. Later, Algiers tried to mediate between Iraq and Iran to bring the war to a halt. Egypt denounced the war and remained neutral until it reacted to Iraq's request for arms, ammunition, and some other military needs, and to Iran's general offensive on Iraqi territory during the last phase of the war.

The rapprochement of the Iraqi-Jordanian regimes was one of the fruits of the war after Iraq moderated its policies in recognition of its need of Jordan to achieve strategic depth. In the course of the war the United States has responded to the economic and military requests of both Egypt and Sudan, and has responded quickly to Saudi requests.

The war presented Israel with a golden opportunity to eliminate its unresolved problems with certain Arab countries and the Palestinian resistance. Israel launched its offensive on Lebanon and assaulted Beirut, then forced most of the Palestinian resistance to leave Lebanon, exploiting the lack of consensus of the Arabs and the ostracism of Egypt from the Arab camp. Israel's joining with Syria and Libya in support of Iran led to a suspension of the Arab Common Defense Agreement.

At present the Iran-Iraq war is the main preoccupation of the Gulf leaders together with the knotty Palestinian issues. The ongoing stalemate on the battlefield holds the everpresent danger of escalation which could endanger the whole area and the superpowers' interests in the region. Such an escalation could always spill over into the Arabian Gulf states. While the impact of the Iran-Iraq war is presently the main preoccupation of Gulf leaders, the unresolved Palestinian issue is seen by all as the main source of tension in the Middle East. The Arab-Israeli problem has a highly physical and emotional impact on most regional actors that results from a genuine sense of solidarity with the Palestinian cause.

Israel considers that the Iran-Iraq war has furthered its interests. The longer the war between the enemies of Israel goes on, the greater will be the casualties on both sides and the more complete the destruction of their economics. One result of the war is the exclusion of Iraq from the Arab eastern front for quite a long time, a factor helping Israel to achieve its goals and objectives in the region. Israel seized a precious opportunity to draw breath while Gulf leaders were preoccupied by events in their immediate area.

Israel has sold American military arms and equipment to Iran, raising the combat efficiency of the Iranian armed forces, and enabling them to continue their military efforts. During the Baghdad air raids, Israel launched its own strike against Iraq's nuclear reactor. Aware that the Gulf states fear an Iranian threat, Israel hoped that these states would seek a resolution of their problems with Israel before the Iranian dangers became imminent. It is in Israel's interest for the war to continue. Israel is well aware that the Iranian-Israeli cooperation is only a temporary arrangement that is unlikely to outlast the war, and Israel recognizes the probability of Iran joining the Arabs after the war ends.

Egypt has objected, and is still objecting to the Iraq-Iran war, because it involves two neighboring Islamic countries. Egypt is also aware of the war's detrimental effects on the Palestinian question and the peace process in the Middle East. When the war turned to Iran's favor, Egypt supported Iraq and supplied it with some arms and ammunition. The victory of Iran and defeat of Iraq would establish a Shia belt, affiliated to Iran, extending through Iran, Iraq, Syria, to southern Lebanon—the belt that Iran dreams about. Thus the Egyptian arms supplies to Iraq are not looked on as support for the ongoing war in the area, but as the unavoidable commitment of the largest state in the area. Egypt chooses to sell military supplies directly to Iraq and refuses to do so through middlemen because it fears Iran would become the ultimate user of shipments made through a third party. Egypt is trying to avoid disequilibrium of the military balance which could destroy the security of Iraq, the Gulf states, and Arab security in general. Egypt fears that continuing Iranian pressure and supremacy would encourage terrorist groups affiliated with Iran to continue insurgency operations similar to those that occurred in Bahrain and Saudi Arabia.

Besides supporting outside efforts to bring the war to an end, Egypt has made an actual contribution in the Security Council, discussing new ways to end the Iran-Iraq war, and supporting all the United Nations' activities in this endeavor. It has also requested several of the world's powers to halt the war. Egypt feels an obligation to further the cause of peace and to ensure security and stability for the region's states.

The Gulf States Today

A wide diversity of views presently exists among the Gulf states. All share a common perception that an escalation of the war poses severe risks for them. Although each has adjusted to the new realities, none see any benefit from a prolongation of the war.

Saudi Arabia's combat aircraft, combined with early warning from the AWACs, give the Saudis the capability to defend against both a small surprise attack and a sustained attack by the Iranian air force. A surprise attack in force would be difficult to stop entirely. The Saudis have demonstrated their willingness to use their capabilities when threatened directly by Iran. They would also retaliate if attacked by Iran: their ground attack F–5s give them a significant deterrent capability.

A modern air defense system does not mean, however, that Saudi Arabia's oil facilities are immune to attack by sabotage. Offshore oil extraction platforms are vulnerable. Moreover, it is impossible to secure the thousands of miles of pipeline, the dozens of gas-oil separation plants, the hundreds of storage tanks, or even the tankers' berthing areas against sabotage. The targets are simply too numerous. There are almost half a million Shiites in the eastern province, some of whom have an affiliation with Iran and might engage in internal terrorist activities.

Saudi Arabia remains committed financially and politically to Iraq. Very clearly, Saudi Arabia believes an Iranian victory over Iraq would place Saudi interests in the region in serious jeopardy. Saudi loans have fueled the Iraqi armed forces (an estimated $15 billion over four years), and Saudi diplomacy has pursued an embargo on arms to Iran. Nevertheless, the Saudis are not prepared to go as far as Iraq desires in actively discouraging purchases of Iranian petroleum.

Located close to the larger states of the region—Saudi Arabia, Iraq, and Iran—the Kuwaitis are particularly vulnerable to external pressures. Currently, Iran appears most threatening; Kuwait considers itself the first target of Iran among the Gulf's states. Kuwait is a lucrative target for Iran. A majority of Iraq's imported military equipment now comes through Kuwaiti ports. Kuwait's key economic and industrial facilities are highly concentrated in one area and thus are particularly vulnerable to air attack. The Kuwait government also faces difficult internal problems. At present, only approximately 42 percent of Kuwait's population are citizens and nearly a third of these are Shiites. Economic problems have also contributed to Kuwait's difficulties; the collapse of the stock market, reduced oil revenues, the flight of capital, and now the war have all combined to strain the economy.

Bahrain and Qatar are under the defensive wing of Saudi Arabia. The Saudi combat air patrol and AWACS surveillance cover both countries against an Iranian air threat. While neither Bahrain nor Qatar has a significant defense capability, neither do they have large industrial areas that are lucrative targets. The principal threat is sabotage and subversion. The coup plot in Bahrain, which was discovered in December 1981, was clearly orchestrated in Tehran. With Shiites forming a majority of the population, Bahrain remains vulnerable.

The Iran-Iraq war has forced tough choices on the United Arab Emirates (UAE). The UAE is willing to fight if attacked but is essentially defenseless against Iran. It is not covered by the Saudi security umbrella or by the GCC defense agreement and, therefore, must be prepared to defend itself. Since Iran is traditionally the largest trading partner of the UAE, the emirates naturally want to avoid direct Arab-Iranian confrontation. Moreover, some of the smaller sheikdoms share oil facilities with Iran and are against putting Iran in a corner.

With the exception of Musandam, the area at the tip of Oman that controls the southern passage of the Strait of Hormuz, Oman is far from the Gulf war. The Omanis have a small but highly capable air force. Oman's forces, however, are deployed south and west against the threat from South Yemen, and would be of little use against an Iranian attack. In general, the Omanis have reasonable relations with Iran and are careful to do nothing to provoke their larger neighbor.

The Omanis echo the common theme that an end to the war is highly desirable. However, Omanis also offer little hope that the war will end soon. Omanis are particularly skeptical with respect to the prospects for a mediated settlement. They express a clear determination to respond to a direct Iranian attack on Oman, but are hesitant to be drawn into the conflict otherwise. In the event of a worsening of the conflict, they stress the need for concerted GCC action, and specifically Saudi participation, before Oman would consider expanding its role.

Global Repercussions

The conflict between the two superpowers in the Arabian Gulf area is the most important result of the Iran-Iraq war. They have both devised new strategies to support their political and military objectives and to maintain their interests in the Arabian Gulf area and in the Middle East as a whole.

The superpowers' national security objectives in the Middle East and especially in the Gulf merit examination. To reiterate, the United States' chief objectives are: to encourage the peace settlement for the Arab-Israeli conflict which will ensure the continued existence and security of Israel; to secure continuous supplies of oil and gas from the Arabian Gulf area to the United States, its Western allies, and Japan; to limit the Soviet Union's influence and any other pro-Soviet influences in the Arabian Gulf area; to reduce the tensions that could lead to a direct confrontation between the superpowers in the Arabian Gulf area; and to insure regional stability.

The effective US military presence in the region is basically a naval presence, and most of the US Navy presence is outside the Gulf and out of sight. A US carrier battle group is currently serving with the MIDEASTFOR in the Arabian Gulf and is considered symbolic of US intent in the area. The CENTCOM forces (a new command created out of the Rapid Deployment Force) deployed in January 1983. CENTCOM has eighteen cargo ships (loaded with supplies for 10,000 troops) anchored 2,300 miles south of the Strait of Hormuz at Diego Garcia, located roughly in the center of the Indian Ocean.

The Carrier Battle Group is generally at least 500 air miles from the lower Gulf, and the carrier's aircraft could quickly be transferred to any available Gulf state airfield in time of need. The conventional warfare capabilities of the fighter and attack aircraft aboard one carrier could probably neutralize the Iranian ports and airbases. The US frigate/destroyer combatants in the Arabian Gulf itself could probably defend against all but a sustained Iranian attack.

There are no likely contingencies in which the full array of the CENTCOM forces might be needed. If the Gulf war should escalate to the point of US military involvement, a deployment might include several squadrons of US fighter aircraft, additional AWACs and tankers, additional destroyers/frigates for convoy duties and, possibly, a second carrier battle group. There is no need for US ground troops except for security guard duty.

The United States is further enlarging and improving the Diego Garcia naval base, and holding negotiations with some of the region's states to exploit the military facilities of Egypt, Sudan, Somalia, Oman, and Kenya in order to achieve supremacy, or at least some sort of military balance with the USSR. The United States conducts periodical military training in the area in order to gain familiarity with local geographic, hydrographic, and weather conditions. It also exploits the exercise facilities made available in friendly countries to perform joint training in ground, air, and sea exercises. American strategic planners set store by the Gulf states' contribution to their self-defense and the visibility of regional collective security systems. Both US policy and military capabilities are heavily dependent upon cooperation with the region's states and the GCC. Tactical air cover is crucial to any US combat operation in the Gulf and access to regional airbases would be needed in any sustained operation.

The United States is studying and trying to interest its Western allies in the possibility of establishing a joint naval force. Again, US policy is dependent on the cooperation of European allies. Political and military coordination with its NATO allies on this matter is already extensive. The French are reluctant to cooperate, despite their significant naval presence at Djibouti and the fact that their help in the region would considerably reduce the strain on the United States. The British have a small naval presence in the area and could make a positive contribution to regional security.

US policymakers have to determine the order of priority among the regional goals of countering the Soviet threat to the Arabian Gulf states; solving the Arab-Israeli conflict with the cooperation of the rest of the Arab states; and completing the peace process in the Middle East. A stalemate to the war would coincide with the US objective of keeping the pro-Soviet Iraq and the anti-American Íran as weak countries, and strengthen the pro-US Gulf states' positions. It would also be a means to achieve relative stability in the Arabian Gulf region. A stalemate would probably lead to the gradual strengthening of US and Western influence in the area as the former belligerents strove to rebuild their economies. A continuous oil flow across the Strait of Hormuz would be ensured by a stalemate.

A decisive Iraqi victory would be unacceptable to the United States. It would represent a victory of Soviet arms over US arms. It would reinforce the Soviet presence in Iraq, in addition to that in Syria and Libya. It would strengthen the common front of enmity towards Israel with a corollary effect on the peace efforts in the region. It would increase Iraqi influence in the Gulf region, particularly on the Gulf states which presently have conservative pro-US regimes. The defeat of Iran might lead to its division which would not be in accord with US objectives. An Iraqi victory would leave Iran with only one alternative: to seek a shelter with the Soviet Union.

A decisive Iranian victory would be no more acceptable to the United States because of the tremendous enmity of the Iranian Revolution towards the United States, and because Iran would be enabled to export its revolutionary principles in addition to exerting control over the region and being in a position to threaten the Gulf states. The instability in the Arabian Gulf region would continue.

The probability of US military intervention in the Arabian Gulf region in support of one of the belligerents is a remote contingency but not an impossible one in the following circumstances: the closure of the Strait of Hormuz and the cutting off of the oil flow to the West; a complete collapse of the Iraqi military situation that could lead to a decisive victory for the Iranian Revolution; the threat to Iran as a state and the possibility of its division if it were defeated and a decisive Iraqi victory were achieved; and direct Soviet military intervention in the favor of one of the belligerents. An American military intervention need not necessarily be by armed forces, but could encompass large-scale military supply delivery, security information, and increased support from technical experts.

The Soviet Union, for its part, considers that the Middle East area in general and the Arabian Gulf area in particular represent the southern security belt of the USSR, and any US control of this area is a direct threat to Soviet national security. Soviet national security objectives are therefore to secure a suitable ambience to spread its ideology throughout the region the better to impose its political and economic influence and to prove its impact on the region in order to facilitate its participation, with the United States, in formulating a resolution of the Arab-Israeli conflict. Such a resolution would aim to secure the existence of Israel and the rights of the Palestinian people, and the Israeli withdrawal to its international borders before 1967. Further Soviet aims would be to secure the oil and gas flow to the Soviet Union, and to the east European states for the next decade, and to sign agreements with the region's states paving the way to warm water access.

To further these ends, an effective Soviet military presence is sought in the Indian Ocean, the Gulf of Oman, and the Red Sea to achieve an acceptable balance with US naval forces. The Soviets deploy an average of about 25 ships to the Indian Ocean. More than 20 Soviet divisions are deployed on the Soviet side of the joint border with Iran. Taking advantage of the Kuwaiti's military requirements the Soviets have recently signed an arms sales agreement with them, thus expanding into the Gulf Cooperation Council states. There will likely be further Soviet efforts to establish relations with the conservative Arab states of the Gulf.

The Soviets seek a position of predominant influence with both Iraq and Iran, and still consider northern Iran to be in their sphere of influence. They are wary of any escalation in the Iran-Iraq war that might engage US forces in keeping the Strait of Hormuz open, while any use of US forces against Iran could provide the Soviets with a political opportunity to move into northern Iran. The Soviets, nevertheless, favor a settlement of the conflict which would facilitate the Soviet pursuit of influence in Iraq and Iran simultaneously. The Soviets are thus seeking to mediate between Iraq and Iran and, should such an effort prove successful, the Soviets would gain credibility throughout the region.

The Soviets reinforce liberation movements with their political and diplomatic support, and this, in addition to the supply of arms that they furnish, tends to increase Soviet political and economic influence. The USSR naturally supports the pro-Soviet regimes in the area in order to achieve their military cooperation, and encourages its allies, Cuba and East Germany, to commit their military expertise and military forces to the area hoping, by these means, to identify new opportunities in the Arabian Gulf area and throughout the Middle East. The Soviets have been willing to invest heavily in Syria through arms sales and the direct military involvement of Soviet advisors. Their injection of military might gave Syria the strength and fortitude to challenge the United States and Israel in Lebanon.

A stalemate to the war would coincide with the Soviet Union's objective of having only weak or allied nations along its borders. However, were a decisive Iranian victory to occur, this would increase the probability of the Iraqi regime's collapse, an undesirable outcome for the USSR which has negotiated a friendship and cooperation agreement with Iraq. Such a victory would swell the tide of Islamic resurgence in the direction of the southern Soviet republics which have large Moslem populations.

Open US military interference in favor of one of the belligerents would likely prompt a Soviet military intervention in favor of the other side. Alternatively, the Soviet Union might interfere to hinder the total defeat of one side. The effect of polarization by the two superpowers has been to draw most of the world's nations into either the Western or Eastern orbit.

The nations belonging to NATO and the Warsaw Pact cluster around the two superpowers, but the roles of France and Japan cannot be overlooked. France plays an independent, neutral role according to its objectives in the region. The French economy needs Gulf oil and credits from the region's states (the petrodollars). The Gulf states offer broad market opportunities in which France wants to have a larger share, particularly in arms sales. France also wants to contribute to the rebuilding and development plans for the post-war era. The Arabian Gulf states encourage France to play a role which coincides with their interests, so that France's closer cooperation with the GCC states in the near future is likely. France is inclined to find the no victory-no defeat ending of the Iraq-Iran war acceptable.

Japan, despite its role as a United States' satellite, has its own policy toward the Gulf states which furthers its own political and economical interests in assuring the oil flow from the Arabian Gulf region to Japan, and the exploitation of the burgeoning Gulf states' market for its products. Stabilization of the region would further Japan's interests.

IV

LIKELY FUTURE DEVELOPMENTS

With armed conflict between the two most powerful Gulf states, the Middle East is presently passing through one of its most crucial phases. As a result, within the framework of world power balance, regional security has been imperiled and stability in general threatened. The Iran-Iraq war has occasioned grave economic deterioration, the loss of many lives, and wholesale attrition of abilities and capabilities in the area.

The current war of attrition began with a border incident instigated by Iraq in the second half of September 1980. The conflict, envisioned as a short, limited, local war, did not achieve its objective of imposing Iraq's will over Iran. The Iraqi offensive against Iran united the Iranian people in demonstrable support of their armed forces against the external danger. With improved morale, they stood firm against the Iraqi offensive.

The dependence of both belligerents on arms and spare parts from foreign countries meant that the supplier nations firmly controlled the events of the war. This permitted its prolongation and the exhaustion of the opposing countries in the interest of the arms sellers. Neither belligerent could regain combat efficiency under these terms.

After evaluating the situation, Iraq determined political and political-military objectives appropriate to its military capability, but failed to achieve these objectives. In a similar way, the Iranian political and political-military objectives remained beyond its capabilities. The termination of the war with no victory and no defeat would harmonize the interests of the Arabian Gulf states and the two superpowers. This would make possible the imposition of an acceptable settlement when the war ends.

On both sides the art of strategy and the art of operation were not always applied properly, so the fighting went on in a systematic style combined with random actions and inaccurate forward planning. The development of Iraqi combat operations aimed, at first, to capture the centers of oil production, the source of Iran's wealth, but this was not accomplished.

The deteriorating domestic situation and the multitude of internal political squabbles in both countries hampered war planning. In Iran, political chaos existed. The impact of military reversals stunned Iraq while the problems with minorities and the deterioration of the economic situation in both countries imposed additional burdens on both regimes and created potential long-term threats.

By analyzing the results of the ongoing war we can determine the factors that will have impact on the Arabian Gulf region during the coming two decades; that is, up to the end of the 20th century. Locally, the physical and economic rebuilding of the warring countries' infrastructure may take long years after the termination of the war, leaving Iraq and Iran politically, economically, and militarily weak. The advanced industrial countries of the West, who manipulate modern technology, will be the beneficiaries. This war may not be the last between Iraq and Iran. On the contrary, it may mark the beginning of a period of even greater animosity between the two countries.

The war has provoked tension and instability within the Arabian Gulf region where the area's security became the first priority. The enmity between Iraq and Syria caused much embarrassment to the Palestinian revolution and postponed the achievement of its objectives.

But for the Iran-Iraq war, Israel would not have been able to accomplish its greatest national security goal: the destruction of the Iraqi nuclear reactor. As a result of the war, the two major Arab and Islamic military powers, Iraq and Iran, were removed, for a considerable time, from the Arab-Israeli conflict. The Israelis are exploiting Iran's urgent need of spare parts from its Western allies to decrease Iran's support of Palestinian leaders.

From an international point of view, the superpowers speedily reinforced their military presence in the Arabian Gulf region according to their national security objectives, either by direct military presence or through signing friendship and cooperation agreements, and obtaining military facilities in the area. The surge of military activity in the Arabian Gulf area has led to a renewed interest in the procurement of US arms. The latest and best military hardware from the United States and other arms suppliers will have a ready market in the region.

The Soviet Union's policy is such that it encourages the United States and the Western industrial countries to direct part of their military and economic resources to the region (the strategic attrition policy) in the interest of the historic ideological conflict. Conflict of power, regionally and internationally, in the Arabian Gulf region is not surprising in view of history and the increasing strategic importance of the area, but this conflict should always be confined within reasonable bounds. Should the conflict exceed a certain limit, it will endanger the area and may even escalate to an irrevocable super-power confrontation.

Neither superpower believes in regional nationalism; both exert great efforts trying to eliminate this concept. They pay a very high price, politically and economically, in this endeavor. No policy will succeed in eliminating nationalism, because it is rooted in the region's history and practice. The best policy, to my mind, is to live with the facts and to give them due regard in the strategic planning process. This would be the cheapest and most effective policy.

The war has exhausted its objectives, locally, regionally, and internationally. It has reached a point where its continuation is a burden to all the concerned parties, including the superpowers. The war should be brought to an end, and I think it would be possible to achieve an armistice.

It should be very strongly emphasized to US policymakers that no country in the Middle East really believes that the true danger in the area is the Soviet threat. Any country appearing to hold such a view is only taking that position because it wants to engage the interest of the United States. Nevertheless, regional states will really co-operate with the United States only after the main problem of the region is solved: the Arab-Israeli conflict.

Peace in the Middle East area should be the first priority of US policymakers. At the core of this peace is the Palestinian question. If the Palestinian question were solved, peace would prevail and sincere efforts would be concentrated on cooperating with the United States.

ENDNOTES

1. *Shia* means disciples or supporters. The Shia are the disciples of *Ali Ben Abi-Taleb,* who was the cousin and son-in-law of the Prophet Mohamed. He married *Fatma-al-Zahraa,* the prophet's youngest daughter, and they were the parents of Al-Hassan and Al-Hussien. After the death of Ali Ben Abi-Taleb there was a disturbance among Moslems and Al-Hassan and Al-Hussien went to Iraq after they had been promised the support of the people of the area. However, they were betrayed and Al-Hussien and other members of his family were

killed while fighting in a famous place called Karbalaa, Iraq. This incident was the origin of the Shia. The Shia still mourn the betrayal of Al Hassan and Al-Hussien, and are recognized as the supporters of Ali Ben Abi-Taleb and his two sons.

2. *Fatma-al-Zahraa* was the Prophet's youngest daughter and the wife of his cousin, Ali Ben Abi-Taleb. Shortly after her father's death she died while still in her twenties. She is an important figure to Moslems in general and to the Shia in particular.

3. *Fateh* means "opening," and it is sometimes used in Arabic to describe a sudden solution to a tough situation. In addition, it is the title of a *Sora*, or part of the Koran, the Moslem holy book. It is linked to an incident at the beginning of Islam when God promised his Prophet Mohamed the "opening of Mecca" and victory over the city's much more numerous inhabitants. Mecca was taken exactly as promised, and from then on Islam prevailed. *"Fateh"* is used as a synonym for victory over the enemy in a holy war.

4. *Bit-al-Makdes* is Arabic for "Jerusalem." At first Moslems prayed in the direction of Bit-al-Makdes until God, through the Prophet Mohamed, told them to pray facing toward the holy building in Mecca built by the patriarch Abraham and his son Ishmael, the father of all the Arabs. Bit-al-Makdes is the third Moslem holy place, the others being the holy building of Abraham and Ishmael in Mecca, and the Prophet's Mosque in Medina, Saudi Arabia, where the Prophet Mohamed is buried.

The three names of the Iranian counterattacks *Fatma-al-Zahraa, Fateh,* and *Bit-al-Makdes* were carefully chosen for their significance to Islam.

THE EFFECTIVENESS
OF MULTINATIONAL FORCES
IN THE MIDDLE EAST

Brigadier General
Yehuda Bar
Israeli Army

CONTENTS

TABLES

MAPS

CHARTS

APPENDIXES

I

THE EFFECTIVENESS OF INTERNATIONAL FORCES

Politicians tend to appreciate the effective use of international forces as a tool for settling international conflicts. This solution has been a common one in the Middle East conflicts. As a tool the use of international force has many advantages as well as shortcomings. In this paper I shall analyze the application of international forces in the effort to settle the continuing, complicated conflict between Israel and the Arab countries. Since 1948, when Israel became a *national political factor,* until the present, employment of this mechanism has met with success as well as failure.

Nothing is gained by analyzing the effectiveness of international forces in general terms: every case needs to be examined in relation to the particular instance, the military-political situation, and local conditions. The question that must always be asked is what lessons, if any, have been learned from the failures as well as from the successes. If previous mistakes and failures are not to be repeated, we must analyze how effectively lessons learned have been applied subsequently. This paper does not intend to answer all the questions: I shall focus on three cases—one success and two failures—dealing with UN forces in South Lebanon and the multinational forces in Beirut. The failures are examples of the complex problems that result from deploying international forces when not all the relevant factors have been taken into consideration. This paper does not seek or presume to recommend a solution to the Lebanese problem but rather to use Lebanon as a platform in examining the use of international forces, as viewed from an Israeli perspective.

This research paper does not represent Israeli official views; it only represents my own views and ideas.

II

THE NATURE OF THE CONFLICT

The Middle East as a geopolitical unit has, for a very long time, stimulated the attention and the interest of many nations, particularly the superpowers. Some of the reasons for this interest in the region, which has increased in the past few years, can be rather simply explained. In the Middle East one finds a concentration of many nations' political objectives. The superpowers' competition for influence and control is but one example. Other factors relate to the region's oil resources and reserves. The most important reason relates to the political and military instability in the region which has caused many military conflicts, all of which could easily deteriorate and draw the superpowers into direct military conflict. Another characteristic of Middle East conflict is that local conflicts can expand beyond the region and affect other parts of the world. As an example, the energy crisis in 1973 originated in the local war between Israel and both Egypt and Syria but it spread and affected all of the Western world. The memory of 1973 compels the nations of the world, and particularly the Western nations, to seek a political process to reduce the possibility of such a crisis occurring again.

International terrorism, with its source in the Middle East, operates worldwide. It is a by-product of the politically confused situation in this region. The painful impact and the fear of being victims of terrorism motivates free nations to become involved and to seek a political solution to the basic problem—the relations between Israel and the Arab countries. Many nations believe that solving this problem will reduce the potential for conflict in the region. The Arab-Israel conflict is a continuing conflict which, from time to time, breaks out into all-out war. Between the major wars political, economic, and small military struggles continue unabated. From a political-military aspect, the Arab-Israel conflict can be divided into three categories:

The *First Category* includes the wars which Israel fought against one or more Arab countries, which can be conveniently summarized thus:

ISRAEL'S WARS, 1948–1982

	Egypt	Syria	Jordan	Lebanon	Iraq
1948	+	+	+	+	+
1956	+				
1967	+	+	+		+
1973	+	+			+
1982		+		+	

This category is characterized by short wars which can be circumscribed by dates and by the countries involved in the war. The wars ended in clearly defined lines—a fact that simplified the call for a cease fire and created initial conditions for negotiation.

The *Second Category* includes all the small and limited military operations which were, essentially, reacting and retaliating against terrorism. Terrorism is used by some of the Arab countries, particularly by the radical countries, as a tool and a means to continue the war against Israel. This category is marked by indefinite results, a vaguely drawn front and confused lines. The objectives and targets may change from fighting against targets of pure terrorism to retaliation and strikes against countries hosting and supporting terrorism. Anti-terrorism activities have often extended from the region to other parts of the world. When such activities escalate, the danger exists that the situation may deteriorate to an all-out war. The political problem in this category is to define the partners, areas, and lines in the conflict, and the unwillingness of the perpetrators of terrorist acts to negotiate and compromise.

The *Third Category* includes enhanced political economic struggles as part of the general conflict, as in the case of the "Arab ban"*

---

* The "Arab ban" of the late forties was the agreement by all the Arab countries that they would combine to oppose Israel, not only militarily but also by every legal means available.

and the sanctions against international companies conducting business with Israel. Initiatives have been taken to exclude Israel from certain international institutions. Massive economic and financial support is given to hostile countries and organizations. This struggle is hard to define in terms of duration and boundaries. Usually, however, this category is characterized by the usage of legal means as part of an all-out struggle.

International Involvement

International involvement relates primarily to the first category of conflict. This involvement influences the nature of the conflict and the concern of the nations in the Middle East. The intensity of international involvement is directly proportional to the intensity of the conflict. It can involve special activities to end military conflict as well as the intensification and deepening of political influence, and, in some cases, even the acquisition of new friends. For instance, after the defeat of Egypt in the 1948 war, the Egyptian disappointment with the Western world was used by the USSR for penetration into Egypt. Soviet influence endured for almost 20 years, until in 1973 the same process turned the Egyptians away from the USSR as a consequence of the 1973 war. The United States initiated negotiations between Israel and Egypt, and by this means intensified its own political influence on Egypt and eliminated the USSR's sway. International involvement in settling Middle East military conflicts has played a part in almost all the wars.

The course of events is usually the same. First, international pressure is applied to convince the countries involved to cease fire. This action can be taken through formal international institutions like the United Nations (UN), or through unofficial and informal approaches, usually with the mutual agreement of the two superpowers. Then, further pressure is applied to cause the participants involved in the conflict to negotiate an agreement which will result in a cease fire. The final phase is the creation of an international mechanism to support and control the execution of the cease fire agreement.

The international mechanism, whose declared mission is to support the sides involved in the conflict, actually serves and supports the initial political objectives of the countries involved in this mechanism. In most cases, partnership in an international mechanism demonstrates and emphasizes the political interest of the country. Such an international mechanism was represented in every case by a military force from 1948 (the first appearance of such a military force) until 1982 when the deployment of the multinational force in Beirut changed the nature, missions, structure, and authority of such multinational forces.

Types of Multinational Forces

First Case—Observers A group of officers from Western countries under the UN flag (which had been placed along the cease fire lines) agreed in 1949 to a cease fire agreement. Their mission was the supervision of the execution of the cease fire agreement, detection of violations, and reporting the facts and conclusions to the belligerents and to the UN. The Observers also served as chairmen of mutual committees (forums in which local problems were discussed and solved). The effectiveness of this model was strongly influenced by the cooperation of the local countries and existed in most places along the cease fire lines until 1967.

Second Case—Emergency Forces Observers (as in the first model) and organized military units were integrated under the UN flag. This system was first used in Sinai following the retreat of the Israeli forces after the Sinai operation in 1956. The force was placed along the Israeli-Egyptian border in an agreed demilitarized strip which was declared a buffer zone. The mission of this force was to control and to define violations of the agreement and to report the facts to the UN. The force had no authority to prevent violations. The force incidentally included countries which had no political ties with Israel like India, Yugoslavia, and other Soviet-oriented countries. The composition of the force influenced the degree of cooperation between the force and Israel. This model existed until 1967.

Third Case—The United Nations Interim Force in Lebanon (UNIFIL) A complete military organization under UN patronage with responsibility in the agreed area, UNIFIL was placed in South Lebanon in 1978, after the Israeli "Litany Operation," had taken place, in a mission to replace the Israeli forces. UNIFIL had the authority to prevent any armed element from penetrating the given zone.

Fourth Case—Multinational Force This is composed of military units from different countries in support of local governments in order to solve political-military problems. An example for this model is the force in Beirut.

The effectiveness and the degree of success of each of the models have been influenced by the provenance of the flag the mechanism is serving; the nationalities of the participating troops; the validity of the declared mission; and the extent of the cooperation of the local partners.

Conditions for the Deployment of International Forces

An international peace-keeping force, although it is a military force, cannot be evaluated and criticized using the same criteria usually used to measure the effectiveness and success of normal military forces. The mission of an international force acting as peace-keeper and sensitive to political objectives in a region in which countries are hostile creates in any such mechanism a hypersensitivity to many influences. One or many local factors can influence the effectiveness of the international force.

The necessary preconditions for the deployment of an international peace-keeping force are first, the international consent among all the influential nations in the given region to use their influence in order to cause the parties in conflict to consent to negotiate and accept the deployment of an international force. Second, the countries in conflict need to be under sovereign regimes strong enough to control the situation and resistant to any internal and external group's opposition to the actual situation. Third, there must be an honest desire on the part of those parties involved in the conflict to end the conflict either temporarily or permanently, and those parties must be willing to cooperate with the international mechanism. Fourth, the international force should have a clear mandate with defined mission and authority. All these prerequisites must be accepted by the countries in conflict and by the nations participating in this international mechanism. In addition to the international force there should be a separate joint committee to the parties in conflict and the international force. This committee should be used as a tool for clarification and for the solving of routine problems or problems which are not defined in the mandate.

Almost all the factors mentioned above should be clarified prior to the deployment of the force. Confirmation of this precondition is the prime condition for success. In examining the Israel-Lebanon conflict, as a background for analyzing UNIFIL and the multinational force in Beirut, it is necessary to deal with the unique political social structure of Lebanon. In fact, this is the key to understanding the problems and difficulties faced by the Lebanese government. It is noteworthy that the same problems influenced the effectiveness of the International Force in both examples.

III

THE SOCIAL AND POLITICAL STRUCTURE OF LEBANON

Lebanese society is a collection of minorities characterized by ethnic groups. Each aspires to the accumulation of power and positions which will increase their particular group's influence among the other groups and will minimize the power of rival ethnic groups. Power can be measured in terms of armed militias and the amount of area controlled by each group. The current situation is the consequence of a long process which started in 1932 when the last population census took place. This census dictated the balance of power among the different ethnic groups and minorities. Until 1984 many changes were taking place in Lebanese society, the most important being the shifting of power among the ethnic groups. Despite the importance of this profound change there was no change in the ratio of key positions in government. Gradually, the balance of relative power between the Christians, who lost their majority, and the Moslems, who became the largest group, shifted. Although the change was a continuing process it reached crisis proportions in the late 1970s.

CHANGE IN RELATIVE POWER FROM 1932 to 1980

ETHNIC GROUP	1932	1980	KEY POSITION
Maronite (C)	29%	24%	President
Greek Orthodox (C)	10%	9%	Commander of the Army
Greek Catholic (C)	6.3%	5%	
Armenian (C)	7.7%	6%	
Total: Christian (C)	53%	44%	
Shiite (M)	18.2%	30%	Chairman of the Parliament
Sunni (M)	20.8%	19%	Prime Minister
Total: Moslem (M)	39%	49%	
Druze	6.3%	6%	

(C) Christian   (M) Moslem

The question that should be asked is how the stability that endured almost 38 years, from 1932 to 1970, suddenly broke down in turmoil and bloodshed. The answer is that an external force destroyed the sensitive balance and caused the collapse of the whole structure. Two external powers associated with the PLO in 1970 started the process and the Syrians in 1975 gave the final push. The interference of the PLO and Syria contributed to and intensified the hostility among the ethnic minorities. The outsiders split Lebanon apart to achieve their own political and military objectives. The critical year in the recent history of Lebanon is 1970. The massive penetration of the PLO and, in effect, the establishment of a Palestinian country on Lebanese territory, particularly in the South and in the Palestinian camps around the big cities, began to put heavy pressure on the Lebanese government at that time. Nothing had been able to stop the extension, consolidation, and aggregation of areas under PLO control. Later in this process, economic and political institutions were established in competition with the legal Lebanese government. The weak government realized the dangers, but was powerless to prevent the deterioration of the situation.

The next step was a request for Syrian military support to stop the PLO's expansion. Syria accepted the request, and sent forces to support the Lebanese government but, a short time afterwards, the nature of Syrian involvement in the situation changed. Instead of supporting the Lebanese government, Syria supported the PLO against the government. Small clashes between the PLO and the government forces escalated into a civil war which still continues to this time. The Israeli invasion of Lebanon in 1982, and the later involvement of the multinational force in Beirut, complicated the situation. New coalitions were created between ethnic groups with new objectives, a recurrent phenomenon in Lebanon. These ethnic groups, with their clashing objectives, and the changing ratios of relative power in Lebanese society constitute the basic problem of the country.

The Government

Since 1970 the government has not had the power to control events. The sensitive balance among the ethnic groups represented in the government, strong inside conflicts on personal interests, and weak leadership reduced the government's capability to take decisions or to react to the events that have led to the present situation.

In 1983, following long discussions and negotiations, a broadly-based government was established with representation of all the ethnic groups. However, participating in the government does not necessarily guarantee a cease fire or the end of hostilities. Alas, the new government did not bring new vigor to Lebanon; it suffered from the same weakness as former governments. The Syrian presence contributed to keeping the government in a weak position. A strong government capable of reuniting the country would have been an obstacle to the Syrian's achievement of their objectives. The Israeli military operations against the PLO and the Israeli Defense Forces' (IDF) presence in Beirut in 1982 did not help support the Lebanese government.

The Armed Forces

The Lebanese armed forces suffered from the same weakness as the rest of Lebanese society. The social shock injured the armed forces so much that it collapsed as an establishment. The Lebanese army ceased to exist as a viable military force during the civil war

(1975–1976) and it has not been a viable force since that time. In fact, the Lebanese army has been the only armed element that has not had combat experience. This state of affairs will continue as long as the central government is weak. The sectoral tensions and the ethnic loyalties within the army outweigh the sense of higher national interest. No change is likely to occur until the independent ethnic armed militias have been disarmed and are no longer more powerful than the army in their respective regions. The Lebanese army pays salaries to about 33,000 soldiers; however, only 20,000 soldiers are at the disposal of the central Lebanese government. Of the 11 brigades in the Lebanese Army (Appendix B), only 5 are available for deployment by the government.

The fighting among the Lebanese in Beirut and western portions of the Shouf proved that the Lebanese army could not function effectively as the military arm of the central government. Inevitably, the army broke down into ethnic components; the debacle was not of a purely military nature but, rather, was based on political/ethnic complications. Most of the army is deployed in camps. Political and ethnic constraints dictate the geographical deployment of the troops. The Shiite part of the army is positioned in West Beirut. The Christian soldiers are located in the eastern part of the capital as well as in the Christian heartland, while the Druze are in the Shouf district.

The Falangs

The Falangs is a political semi-military organization which includes the majority of the Christian population in and around Beirut. This military militia of about 15,000 armed people is supported by some tanks and a few artillery pieces. The Christian community is defended by the military spread out in the eastern part of Beirut, along the beach north of the city, and in the high ground (known as the Lebanon mountain area) which controls the main road to Syria. The organization is led by its creators, the Gemayel family, whose members have furnished two presidents of Lebanon. The first, Bachier, came into power in 1982 and, after a short time, was assassinated. His brother, Amin, is the present president. The objectives of the Christian organization are the preservation of the present situation in which they hold the most important, powerful positions in the government; reunification of the country; and the removal of Syrian and PLO forces from Lebanon. The Christian military militia carry out the main fighting against the Syrian and PLO forces and now they are fighting against the Druze and El Amal as well. In some cases, the Falangs are ready to cooperate with Israel in solving the Lebanese problem.

The Druze

"The Lebanese National Movement" in the past was the organization of all the left wing parties in Lebanon. Since 1982, disagreements between participating parties depleted the organization which then became a purely Druze movement. This ethnic group, like the others, has an armed militia of approximately 12,000 supported by some tanks and artillery. The main weapons suppliers are the Syrians, who are also the group's main political supporters. The Druze militia are located in the Shouf zone east of Beirut, and in the Matan, northeast of Beirut. The Matan area is controlled by the Syrians. The political objectives of the Druze are the preservation of the Shouf district as pure Druze territory; and the achievement of a power position in the government corresponding to their population numbers. The achievement of both objectives will necessitate overcoming their main opponents, the Christians, represented by the Falangs and the Lebanese army. For the Druze any governmental organization symbolizes the Christian faction. The Druze have collaborated with other ethnic groups like the Shiite in fighting the Christians, but there is no formal cooperation between the Druze, as an ethnic group, with the PLO.

El Amal

El Amal is a Shiite religious organization created in 1970 by the Himam Musa Sader with the goal of taking care of the undeveloped Shiite population, particularly in backward South Lebanon. Later, the organization added two additional goals of protecting the Shiite population and advancing the political rights of their ethnic group. To achieve the new goal an armed militia was created and the leadership moved from South Lebanon into West Beirut from where they controlled their operations and activities.

Influences from the Iranian revolution and the acceptance of orders, direction, money, and weapons from Iran has made El Amal a radical religious movement. Groups from this movement are known to have conducted terrorist activities both in and outside Lebanon. The movement is an enemy of all non-Moslem groups and particularly of any foreign force whether they be Israelis, UNIFIL, or multinational forces. El Amal, which strongly influences and controls the population, is the greatest obstacle to settling the problem of Lebanon.

IV

SYRIA

The Syrian presence in Lebanon poses a persistent danger to the Christian Lebanese government as well as serving to shield PLO terrorist activities against Israel and against the Lebanese forces. Syrian forces first entered Lebanon in 1976 under the auspices of the "Inter-Arab Deterrence Force," which was invited by the Lebanese government to act as a buffer between the warring factions in the civil war. The Syrians' motives and objectives in Lebanon soon became clear. The Syrian military presence was exploited by Damascus in an attempt to achieve the vision of a "greater Syria" which would include Lebanese territory. Damascus viewed herself as the master of the rest of Lebanon as well.

The Syrian army had to fight bloody battles, primarily against the Falang Lebanese Christian forces who were the only element standing in the way of Syrian hegemony. The Lebanese-based Syrian army posed the danger that the Lebanese government would fall under complete Syrian domination and that the PLO, encouraged by Syria, would seize control of Christian areas and enlarge its influence and control in the rest of Lebanon. In addition to the PLO, Syria controlled and supported two other groups, the Druze who fought under the Syrian flag against the Christian Falangs, and the pro-Syrian Faranjia Christian militia, based around Tripoli. This group is opposed to other Christian camps.

VARIOUS MILITIAS OPERATING IN LEBANON

ORGANIZATION	POLITICAL OBJECTIVES	MILITIA (NUMBERS)	SUPPORTED BY	OPPOSED TO
Christian "Falangs"	A/Preservation of the Christian superiority B/Unification of the country C/Removal of the Syrians D/Removal of the PLO and the Palestinians	15,000	Central government	Druze, El Amal, PLO, Syria, Faranjia
Druze Lebanese National Movement	A/Preservation of the Shouf District as Druze B/Improvement of the Druze status in the government	12,000	Syria	Falangs, Lebanese Armed Forces
Shiite El Amal	A/Improvement of the Shiite's political status B/Removal of foreign forces from Lebanon	10,000	Iran, Syria	Falangs, UNIFIL, Israel, Multinational Forces
PLO	A/Creation of a "PLO country" in Lebanon	10,000	Syria, other countries	Falangs, UNIFIL, Lebanese Armed Forces
Faranjia	A/Pro-Syrian country	2,000	Syria	Falangs
Syria	A/Dominion of Lebanon as part of Greater Syria	80,000	——	Falangs, Lebanese Armed Forces

POWER CENTERS AND SPHERES OF CONTROL IN LEBANON

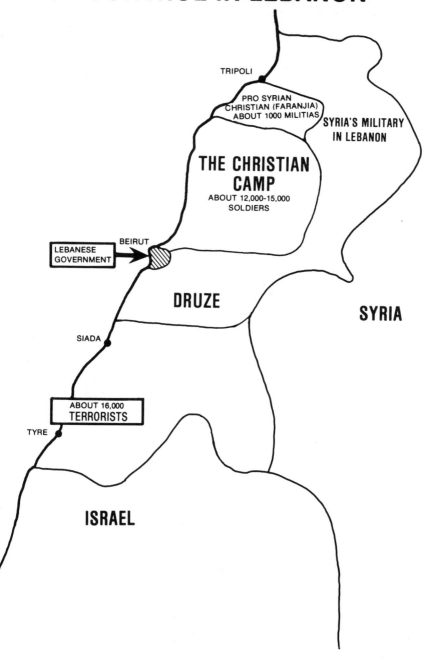

V

THE UNITED NATIONS INTERIM FORCES IN LEBANON (UNIFIL)

In late 1970 the PLO faced the fundamental problem of their very existence. King Hussein drove them from his kingdom. Syria and Egypt restricted the possibility of operation from their countries; the only available territory remaining was Lebanon which offered the best political and geographical conditions. A weak government and armed forces, the consolidation of Palestinians in refugee camps, an appropriate terrain, a short-range from the Israelis' sensitive targets, and a long coastline for sea operations made Lebanon a classic substitute for Jordan.

From 1971 until 1978 the PLO actively conducted armed operations from South Lebanon against Israel, particularly against the northern part of Israel, by shelling civilian settlements with long-range artillery and various kinds of rockets. The seven years of terrorism caused heavy casualties, especially among civilians, women, and children. This intolerable situation gave Israel no choice. Lebanon refused to accept responsibility for the events in South Lebanon or along the border, and lacked the capability to take control of the area. The only option understood by the PLO and the Lebanese was the military option. Retaliation against PLO terrorist activities solved the problem for a short time.

Between 1970 and 1978 the Israeli Defense Forces (IDF) conducted innumerable operations at various intensities and with various objectives, from small, short-range militia operations to full-scale operations. In 1978, the "Litany Operation" was mounted with the objectives of destroying the PLO infrastructure in South Lebanon and creating a security zone along the border as a long-term solution. The idea was based on the conviction that the local population of South Lebanon, without the PLO, would serve their own interest as well as the security of Israel. This optional solution was influenced by previous experience of the government of Lebanon giving proof that it lacked the capability to control this part of the country. This failure

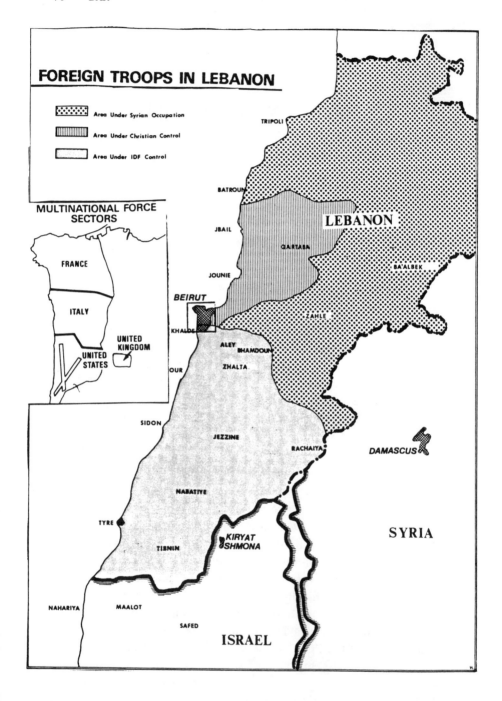

FOREIGN TROOPS IN LEBANON

Area Under Syrian Occupation

Area Under Christian Control

Area Under IDF Control

MULTINATIONAL FORCE
SECTORS

FRANCE

ITALY

UNITED
KINGDOM

UNITED
STATES

TRIPOLI

BATROUN

JBAIL

QARTABA

JOUNIE

BEIRUT

KHALDE

LEBANON

BA'ALBEK

ZAHLE

ALEY
BHAMDOUN

OUR

ZHALTA

SIDON

JEZZINE

RACHAIYA

DAMASCUS

NABATIYE

TYRE

TIBNIN

KIRYAT
SHMONA

SYRIA

NAHARIYA

MAALOT

SAFED

ISRAEL

of control was acknowledged in a letter from the Lebanese ambassador to the president of the UN Security Council, "It is a well known fact that Lebanon is not responsible for the presence of Palestinian bases in Southern Lebanon." The government of Lebanon, which demonstrated and acknowledged its weakness and was powerless to control South Lebanon, rejected any negotiation with Israel or any solution to the area's problems that involved local forces supported by Israel. The Lebanese request for international involvement was a more convenient solution, although this was not a solution favored by Israel.

The Character of the Conflict

In this case the conflict was apparently between Israel and Lebanon but actually other parties were involved, like the PLO and Syria, who were opposed to the achievement of an agreement between Israel and Lebanon. The four specific positions are as follows:

Israel — Is interested in securing the defense of its country. To achieve this goal Israel needs to prevent the PLO from reorganizing and controlling South Lebanon. Israel accepted the Lebanese ownership of the area and was ready to cooperate and support Lebanon in the enforcement of law and order in the area. On the other hand, Israel has rejected the PLO as partners in any negotiation or as partners in an agreement.

Lebanon — On one hand, is not capable of controlling the given area to enforce law and order and to prevent the PLO from dominating the area. On the other hand, Lebanon is not willing and ready to enter negotiations with Israel in order to find a mutual solution. Internal political confusion creates difficulties in the consolidation of an unequivocal position. The request for international involvement is actually an attempt to defer the acceptance of a practical solution.

PLO — The real instigator of the conflict is interested in restoring the previous situation in which the PLO dominated the area. Any agreement between Israel and Lebanon will prevent the PLO from reestablishing its influence in the area. An international

involvement as a temporary solution has high
sensitivities to political pressures and certainly an international force will have less motivation to react
against the PLO than Israel. Therefore, for the PLO,
international involvement is the least troublesome
option.

Syria   — Is not a direct partner in any negotiation or agreement
but indirectly plays an important role in the conflict.
Syria supports the PLO, so it will be unlikely to accept any agreement that restricts the activities of the
PLO.

The disagreements and divergent basic options were apparent
prior to the UN resolution of deployment. UN forces in South
Lebanon continue to exist even after application of the resolution for
the deployment of UNIFIL in the region.

The UNIFIL Mandate

Because of the complexity and the sensitivity in the formulation
and preparation of such documents as the UNIFIL Mandate, I have
included the original document in Appendix A. It is important to understand the general intent of the document. To simplify the criteria
and to evaluate the implementation of the mandate the lengthy document can be roughly summarized under four main headings:

- Retirement of Israeli forces from South Lebanon.
- Rehabilitation of peace and the security in South Lebanon.
- Demilitarization of the area and prevention of any armed element
from entering the UNIFIL area.
- Support for the Lebanese government in restoring law and con-
trol of the given area.

Implementation of the UNIFIL Mandate

Article A in the UNIFIL mandate dealing with the Israeli withdrawal was fully accomplished. The UN forces occupied the positions and the area evacuated by the IDF. The implementation of the
other three articles, which are less technical but more essential, has
never been accomplished. Many factors, some foreseen and others
quite unexpected, combined to cause the declared mission to fail. The

first factor was the well-known fact of the weakness of the Lebanese government and armed forces and their inability to seize the opportunity to take over control of this part of the country. This fact put UNIFIL in a unique situation. Instead of being a supporting force it became the main force, with responsibility for part of Lebanon. Although the central government declared on frequent occasions that South Lebanon is part of Lebanon, at the same time the government blamed Israel for its annexation of the South. The government, in fact, did not cooperate with UNIFIL to demonstrate and exercise its authority either in the area or over the population. This fact eliminated any chance of accomplishing article D of the mandate.

The other cause of failure stems from the fact that the PLO was not a partner in the negotiation. Furthermore, the PLO tried every available means, political as well as military, to prevent UNIFIL from carrying out the mandate. This situation particularly influenced the implementation of article C. A practical interpretation of article C in the given situation implies the use of sufficient military strength—with all the implications and risks involved in that use. This problem raises some questions: first, what motivation had the force to be involved in active fighting? The second, and more important question is: to what extent were the countries represented in UNIFIL expected to sacrifice their soldiers' lives for UNIFIL goals?

The situation in the area provided low motivation for force units fighting against the PLO. The will of the units was affected by the strong resistance to UNIFIL shown by the PLO which very early caused large numbers of casualties. Motivation deteriorated to the point where military missions were avoided in the interests of self-preservation—a situation that enabled the PLO to achieve its main objective to recreate and organize operational bases in the area.

A factor which even more strongly influenced the implementation of article C derived from the insufficient support and backing from the member nations of UNIFIL. Gradually, more and more countries recognized the PLO as a legal political element, putting the member nations in a difficult position. On one hand they recognized the PLO, on the other hand they had to fight this organization. Accordingly, in most events the decision was made to reduce the volume of activity.

MILITARY ORGANIZATION AND CHAIN OF COMMAND

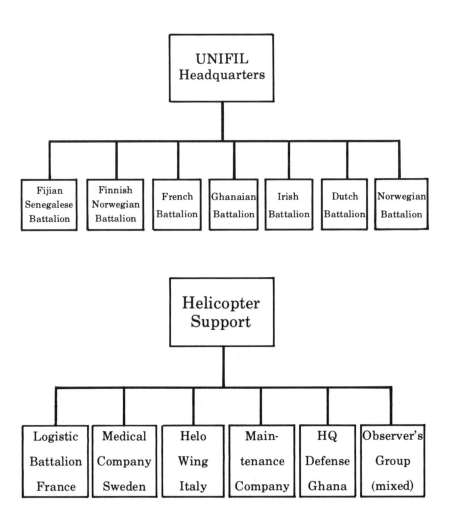

TOTAL: 6000 Men

Has UNIFIL Accomplished its Mission?

The answer is a decided "no!" The clearest proof is the PLO's ability to use the UNIFIL area as a firm base for the renewal of hostilities against Israel. The terrorists' activities have taken place in daylight without fear of interruption or retribution. In the few cases where members of armed PLO groups were caught, the most common result has been the release of the group. In many cases during subsequent negotiations different factors combined to reduce the effectiveness of the force to the lowest degree. The following question should be asked: Was there any chance the UNIFIL would be able to accomplish its mission? Again, the answer is "no." Almost all the essential conditions necessary for success were unfulfilled. An unclear mandate with a complicated military structure, and ties with outside political pressure that were created by nations participating in the peace-keeping force in order to achieve their own objectives, negated any possibility of success.

The problem becomes more complicated when we analyze the position and the capabilities of the parties to the conflict. On one side, Israel, with a sovereign government, had a real desire to settle the conflict in a way that would ensure the security of the northern part of its country. On the other side, Lebanon, which should have played the principal part in cooperating with UNIFIL, was represented by a weak government, so limited by internal difficulties that any action taken in the south could be projected in Beirut and used to increase the opposition to the government. The other two parties in the conflict, Syria and the PLO, were not partners in the agreement. Furthermore, Syria and the PLO both rejected the UN decision. Although all these facts were known in advance of the deployment of UNIFIL in the area, the Lebanese government's inability to cooperate with the force created a situation in which UNIFIL became responsible for an area instead of being dedicated to the original mission of peacekeeping. UNIFIL was not capable of performing the mission it acquired by default.

The Background to the Multinational Force Deployment in Beirut

In the summer of 1982 a very critical situation developed in Lebanon, especially around Beirut. The IDF captured and controlled

South Lebanon, and Beirut was surrounded by Israeli forces combined with the Christian militias. In the western part of the city three armed elements were concentrated—the Syrian 85th Infantry Brigade; approximately 15,000 members of the PLO (part of the force withdrew from the south where some had remained from the beginning including the headquarters and the leaders of the organization); and a third element, a Moslem armed militia, the "Morabitun," under Syrian control. The civilian population of over 1 million suffered casualties and lacked the means of existence.

Israel claimed to have accomplished the primary objective of the war, "the destruction of the terrorist infrastructure in Lebanon." (The PLO used Beirut as a training, logistic, and departure base). It also claimed to have made possible the development of peace talks between Israel and Lebanon without fear of PLO or Syrian domination. Hope of really achieving these objectives caused Israel to continue the war. The PLO and Syria, for their part, sought to ensure the continued existence of the PLO as an organization in Lebanon; the evacuation of the beseiged Syrian Brigade; and continued Syrian domination of Lebanon.

The situation posed the danger of the revival of the war between Israel and Syria with the risk of escalation and expansion to other sectors. Political negotiation, sponsored by the United States, began at the end of July. On 19 August 1982 the governments of Israel and Lebanon reached an agreement for the withdrawal of the PLO and the Syrians from Beirut out of Lebanon under the supervision and security of a multinational force. The force took up positions on 20 August and on 21 August the PLO and the Syrians began to leave the Lebanese capital. Their expulsion was completed on 1 September. Two days later the multinational force left the area. The only armed militia remaining in Beirut was the "Morabitun," a force of approximately 3,000 armed men. This element continued its resistance to the Lebanese government, and from time to time acted against the IDF. The second phase of the multinational involvement followed the Israelis' entry into West Beirut as a result of and in response to the murder of the newly elected president, Bachier Gemayel. The IDF entered West Beirut in order to prevent disaster and bloodshed, and to enable the election of a new president, Amin Gemayel, to proceed. The Lebanese government, under the new president, requested international support to control the situation, in the belief that such involvement would force Israel to withdraw from the western part of the city and maybe from all Lebanon. The following are the Lebanese request and the US reply.

The Lebanese Request for Deployment of a US Force to the Beirut Area

Your Excellency: I have the honor to refer to the urgent discussions between representatives of our two governments concerning the recent tragic events which have occurred in the Beirut area, and to consultations between my Government and the Secretary General of the United Nations pursuant to United Nations Security Council Resolution 521. On behalf of the Republic of Lebanon, I wish to inform your Excellency's Government of the determination of the Government of Lebanon to restore its sovereignty and authority over the Beirut area and thereby to assure the safety of persons in the area and bring an end to violence that has recurred. To this end, Israel forces will withdraw from the Beirut area.

In its consultations with the Secretary General, the Government of Lebanon has noted that the urgency of the situation required immediate action, and the Government of Lebanon, therefore, is in conformity with the objectives in UN Security Council Resolution 521, proposing to several nations that they contribute forces to serve as a temporary Multinational Force (MNF) in the Beirut area. The mandate of the MNF will be to provide an interposition force at agreed locations and thereby provide the multinational presence requested by the Lebanese Government to assist it and the Lebanese Armed Forces (LAF) in the Beirut area. This presence will facilitate the restoration of Lebanese Government sovereignty and authority over the Beirut area, and thereby further efforts of my Government to assure the safety of persons in the area and bring to an end the violence which has tragically recurred. The MNF may undertake other functions only by mutual agreement.

In the foregoing context, I have the honor to propose that the United States of America deploy a force of approximately 1,200 personnel to Beirut, subject to the following terms and conditions:

—The American military force shall carry out appropriate activities consistent with the mandate of the MNF.

—Command authority over the American force will be exercised exclusively by the US Government through existing American military channels.

—The LAF and MNF will form a Liaison and Coordination Committee, composed of representatives of the MNF participating governments and chaired by the representatives of any Government. The Liaison and Coordination Committee will have two essential components: (A) Supervisory Liaison; and (B) Military and technical liaison and coordination.

—The American force will operate in close coordination with the LAF. To assure effective coordination with the LAF, the American force will assign liaison officers to the LAF and the Government of Lebanon will assign liaison officers to the American force.

The LAF liaison officers to the American force will, *inter alia,* perform liaison with the civilian population, and with the UN observers and manifest the authority of the Lebanese Government in all appropriate situations. The American force will provide security for LAF personnel operating with the US contingent.

—In carrying out its mission, the American force will not engage in combat. It may, however, exercise the right of self-defense.

—It is understood that the presence of the American force will be needed only for a limited period to meet the urgent requirements posed by the current situation. The MNF contributors and the Government of Lebanon will consult fully concerning the duration of the MNF presence. Arrangements for the departure of the MNF will be the subject of special consultations between the Government of Lebanon and the MNF participating governments. The American force will depart Lebanon upon any request of the Government of Lebanon or upon the decision of the President of the United States.

—The Government of Lebanon and the LAF will take all measures necessary to ensure the protection of the American force's personnel, to include securing assurance from all armed elements not now under the authority of the Lebanese

Government that they will refrain from hostilities and not interfere with any activities of the MNF.

—The American force will enjoy both the degree of freedom of movement and the right to undertake those activities deemed necessary for the performance of its mission for the support of its personnel. Accordingly, it shall enjoy the privileges and immunities accorded the administrative and technical staff of the American Embassy in Beirut, and shall be exempt from immigration and customs requirements, and restrictions on entering or departing Lebanon. Personnel, property, and equipment of the American force introduced into Lebanon shall be exempt from any form of tax, duty, charge, or levy.

I have the further honor to propose, if the foregoing is acceptable to your Excellency's Government that your Excellency's reply to that effect, together with this note, shall constitute an agreement between our two Governments.

Please accept, Your Excellency, the assurances of my highest consideration.

/s/Fouad Boutros
Deputy Prime Minister/Minister of Foreign Affairs
September 25, 1982.

The US Answer to the Request

Your Excellency: I have the honor to refer to your Excellency's note of 25 September 1982 requesting the deployment of an American force to the Beirut area. I am pleased to inform you on behalf of my Government that the United States is prepared to deploy temporarily a force of approximately 1,200 personnel as part of a Multinational Force (MNF) to establish an environment which will permit the Lebanese Armed Forces (LAF) to carry out their responsibilities in the Beirut area. It is understood that the presence of American force will facilitate the restoration of Lebanese Government sovereignty and authority over the Beirut area, an objective which is fully shared by my Government, and thereby further efforts of the Government of Lebanon to assure the safety of persons in the area and bring to an end the violence which has tragically recurred.

I have the further honor to inform you that my Government accepts the terms and conditions concerning the presence of the

American force in the Beirut area as set forth in your note, and that Your Excellency's note and this reply accordingly constituent an agreement between our two Governments.

/s/Robert Dillion
US Ambassador

This document later became the draft copy for the agreement between the two governments for deploying the multinational force in Beirut.

VI

IMPLEMENTATION OF THE AGREEMENT

Before examining the actual implementation of the agreement it is essential to analyze the document as the basis for understanding the events that occurred while the force was in Beirut. The document should be analyzed through the following components: the military mission, and the authority to accomplish that mission; the Lebanese government and armed forces' capacity to carry out their part in the agreement; and the wider objectives which the United States sought to achieve.

The Military Mission

In one of the JCS documents, the mission is summarized as follows:

A.   To help establish a stable, secure environment in Beirut.
B.   To enable the government of Lebanon to extend its legitimate authority within the sovereign territory of Lebanon.

The authority of the force was indicated in the same document as follows:

The multinational force's role is one of an interposition force, and it has been directed to follow peace-time rules of engagement. . . . The multinational force is authorized to take only the necessary action to assure its own safety. . . . The concept of operation calls for the force to report violations of se-

security to Lebanese authorities who are to take the necessary action.

The force's mission, as it can be understood from this document and from talking with people involved in defining the mission, can be compared to a police function, with activities characterized by clear rules accepted on both sides with the backing of law and the power to enforce the law. In this case, the actual situation was quite otherwise. The force was considered by some elements as an enemy, and every element or faction involved accepted a different rule and a different law. A more critical and decisive factor was the fact that the legal government supposed to enforce the law had no power to enforce its legal authority. In such conditions the multinational force was unable to operate to achieve the original mission.

The Lebanese Government

To reiterate, the Lebanese government and its armed forces, who were supposed to have the chief responsibility for maintaining the situation, were not qualified or capable of carrying this heavy burden. This fact, although it was known before the multinational force entered Beirut, had not been considered in defining the mission. Even later, after the situation became clear and, in fact, the actual mission became a security force, the inadequacies of the Lebanese were still not adequately addressed.

US Political Objectives

The JCS emphasized that, "The US participation in the multinational force supports such US objectives as the removal of all foreign forces from Lebanon, extension of government of Lebanon's authority throughout Lebanon and the guarantee of security for the Israeli's northern border." In this statement, I think, lay the real objective of US intervention in Lebanon. The foreign forces, Israeli and Syrian, have opposing interests in Lebanon. I estimate that the policymakers in the United States had the feeling that the situation in Lebanon could be used as an initiative for new negotiations and the renewal of peace initiatives. This estimation is reinforced by the mediation efforts of Philip Habib. Success or even partial movement might have increased US prestige, would have solved the Lebanese problem, and would have opened new approaches in other sectors.

VII

UN DISENGAGEMENT OBSERVER FORCE IN THE GOLAN HEIGHTS (UNDOF)

In late 1973 a new military situation was created on the Israeli-Syrian front. The 1973 war ended after the Israeli success in pushing back the Syrian forces to behind the 1967 cease fire line. Counterattacks brought the Israeli forces close to the Syrian capital, Damascus, while destroying the majority of the Syrian army. This situation created a direct threat to the Syrian capital and the Syrian regime. International efforts and pressures caused Israel and Syria to accept a cease fire agreement signed 24 October 1973. The goal of this agreement was to freeze the actual situation, in order to prepare the background for a future comprehensive and fundamental disengagement agreement. The cease fire agreement of 24 October did not last long; indeed the movement stopped, but, on the other hand, the war had changed to a static war of attrition.

The rapid recovery of the Syrian army, with intensive Soviet resupply and numerous Soviet advisers, encouraged the Syrians to resume military activities. The entire situation created the risk of a renewed, full-scale war between Israel and Syria. The superpowers were still under the gloomy impression of the grave crisis between them in mid-October when the Middle East war created the risk of direct military conflict between the United States and the Soviet Union. So great was the risk that nuclear readiness had been declared. In these conditions, the superpowers spared no effort to bring about a disengagement at the front in order to reduce friction and reduce the risk of renewed full-scale war. The United States carried much of the responsibility in preparing the parties and creating a suitable background environment for bringing Israel and Syria to the negotiation table in Geneva under UN patronage. On 31 May 1974 a disengagement agreement had been signed. This agreement brought 11 years of relative relaxation of tension along the border. In order to secure the agreement both sides agreed to the deployment of an international force under the UN Flag.

UNDOF—Military Structure and Mission

UNDOF missions, according to the disengagement agreement and UN documents, involved supervision of the cease fire agreement; arranging for the inspection of disengagement and the dilution of forces and combat means; and supervision of the existence of the agreement. The force of 1,400 personnel was organized thus:

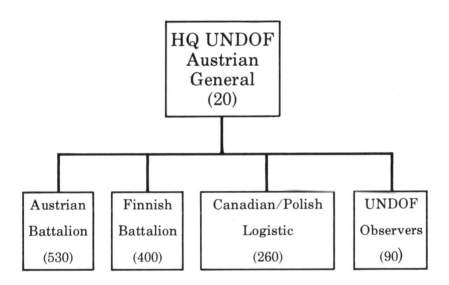

The Implementation

Despite the deep hostility between Israel and Syria in general, it can be said that the agreement was implemented without significant deviation from the original agreement. The system of biweekly inspections on both sides gives the parties confidence.

The reason why this agreement has been so successful and has existed for more than 10 years without major violations is related to the following factors:

—Both Israel and Syria are sovereign governments with effective control of their military and other components in their countries.

—Syria, which sponsors, supports, and encourages the PLO to operate against Israel, has particularly close control of this organization and is able to prevent them from operating outside the terms of the agreement.

—Both Israel and Syria, for their own reasons, are interested in implementing the agreement. Therefore, both countries cooperate with the UNDOF and support the force's mission.

—The strong consensus between the superpowers and other nations that a breach or disruption of the agreement could lead to another military conflict, with the risk of international involvement, makes the superpowers especially sensitive to the issues. They are impelled to keep an eye open on this area and to oversee the implementation of the agreement.

—UNDOF fulfills its mission in an objective fashion without too much involvement of the countries participating in the force.

In sum, the primary reason for UNDOF's success is that in this case almost all the preconditions and actual conditions have been met, enabling UNDOF to operate and to accomplish its missions.

VIII

CONDITIONS FOR AN EFFECTIVE MULTINATIONAL FORCE

A multinational force cannot solve conflicts alone. Its effectiveness is impeded by integration with political campaigns and the outcome can be influenced by the partners' strong desire to reach a solution. Hostility and the combination of emotional and concrete factors create difficulties or, in some cases, prevent direct negotiations. The international method represented by multinational military force acts as initiator of the negotiating process which later on creates agreements.

Therefore, it would be an elementary mistake to evaluate the success of such a force in military terms and according to standard military criteria. The only way to evaluate the force is to consider it as a political tool. As such the force should be limited by the agreement or by its mandate and controlled by civilian politicians. A dangerous situation is created when, contrary to the agreement, the role of the force is shifted from that of supporter to enforcer of the agreement. When this occurs, the force is put in the position of being a partner to the conflict. This, it seems to me, is what happened in both cases in Lebanon.

The Middle East is characterized by an uncompromising situation—many years of hostility, contrasting interests, radical powers, and psychological barriers. All these factors, present in the past, still preclude frank discussion between opposing sides today.

The comparison between the three examples described in this paper can clarify and emphasize the causes of failure in the two Lebanese cases and the success of the Syrian case.

In the Lebanese case, I would claim that not one of the fundamental, necessary preconditions to the deployment of international force had been fulfilled. In addition, in the area of operation, the indigenous countries were lacking in will and the legitimate government was unable to control the interior forces and to cooperate with the international force in settling the problem. The third case presents the opposite situation and can be used as an example of success that precluded any outbreak of hostilities. Success or failure does not derive only from the military capability of the multinational force in carrying out its missions. The main consideration is the countries' willingness and motivation to preserve and execute the agreement and the degree of their cooperation with the international institutes. In the Syrian case, two strong governments were involved, with capabilities to control every element in the area, and they used every means to protect the agreement. The same Syria which gave the PLO their patronage and encouraged this organization in its activities against Israel in other sectors, such as Lebanon, succeeded fully in preventing the PLO from operating in the Golan Heights. One of the main objectives of Syrian policy is securing the situation on their border with Israel.

Another example of success can be taken from the situation between Israel and Egypt. The peace agreement was signed with US mediation. Both sides agreed to open the peace process, and this necessitated the positioning of a multinational force to uphold the agreement until both countries could trust one another and give proof of a real will to peace. The intermingling of internal and external political problems, and the involvement of additional forces that not only were not part of the negotiations but tried to prevent the negotiations in order to protect their own objectives in this country all played a part in the failure of the multinational force in Lebanon. In face of the mainly weak governments and the continuing armed struggles among the ethnic groups, perhaps the only solution can come about after a Soviet determination to force the Syrians to accept negotiations with Lebanon and possibly with Israel.

In any case, with so many political and military conflicts in the world, there should be an international mechanism capable of giving support to countries in solving problems. The only available mechanism seems to be the international institutions like the UN, or other combinations of nations. This method and involvement, political or military, requires careful pre-planning and the creation of sufficient conditions for success before the force is put in place.

Failure of international involvement damages not only the participants' countries but prolongs the conflict and creates a feeling of mistrust in the international establishment.

APPENDIX A

UNITED NATIONS SECURITY COUNCIL
Annex A
AGREEMENT ON DISENGAGEMENT BETWEEN ISRAELI AND SYRIAN FORCES

A.   Israel and Syria will scrupulously observe the cease-fire on land, sea and air and will refrain from all military actions against each other, from the time of the signing of this document, in implementation of United Nations Security Council resolution 338 dated 22 October 1973.

B.   The military forces of Israel and Syria will be separated in accordance with the following principles:

1. All Israeli military forces will be west of the line designated as Line A on the Map attached hereto, except in the Quneitra area, where they will be west of Line A-1.

2. All territory cast of Line A will be under Syrian administration, and Syrian civilians will return to this territory.

3. The area between Line A and the line designated as Line B on the attached Map will be an area of separation. In this area will be stationed the United Nations Disengagement Observer Force established in accordance with the accompanying protocol.

4. All Syrian military forces will be east of the line designated as Line B on the attached Map.

5. There will be two equal areas of limitation in armament and forces, one west of Line A and one east of Line B as agreed upon.

6. Air forces of the two sides will be permitted to operate up to their respective lines without interference from the other side.

C.   In the area between Line A and Line A-1 on the attached Map there shall be no military forces.

D.   This Agreement and the attached Map will be signed by the military representatives of Israel and Syria in Geneva not later than 31 May 1974, in the Egyptian-Israeli Military Working Group of the Geneva Peace Conference under the aegis of the United Nations, after that group has been joined by a Syrian military representative, and with the participation of representatives of the United States and the Soviet Union. The precise delineation of a detailed Map and a plan for the implementation of the disengagement of forces will be worked out by military representatives of Israel and Syria in the Egyptian-Israeli Military Working Group who will agree on the stages of this process. The Military Working Group described above will start their work for this purpose in Geneva under the aegis of the United Nations within 24 hours after the signing of this Agreement. They will complete this task within five days. Disengagement will begin within 24 hours after the completion of the task of the Military Working Group. The process of disengagement will be completed not later than 20 days after it begins.

E.   The provisions of paragraphs A, B and C shall be inspected by personnel of the United Nations comprising the United Nations Disengagement Observer Force under this Agreement.

F.   Within 24 hours after the signing of this Agreement in Geneva all wounded prisoners of war which each side holds of the other as certified by the ICRC will be repatriated. The morning after the completion of the task of the Military Working Group, all remaining prisoners of war will be repatriated.

G.   The bodies of all dead soldiers held by either side will be returned for burial in their respective countries within 10 days after the signing of this Agreement.

H.   This Agreement is not a Peace Agreement. It is a step towards a just and durable peace on the basis of Security Council resolution 338 dated 22 October 1973.

FOR ISRAEL:

_____

FOR SYRIA:

_____

WITNESS FOR THE UNITED NATIONS:

_____

PROTOCOL TO AGREEMENT OF DISENGAGEMENT BETWEEN ISRAELI AND SYRIAN FORCES CONCERNING THE UNITED NATIONS DISENGAGEMENT OBSERVER FORCE

Israel and Syria agree that:

The function of the United Nations Disengagement Observer Force (UNDOF) under the agreement will be to use its best efforts to maintain the ceasefire and to see that it is scrupulously observed. It will supervise the agreement and protocol thereto with regard to the areas of separation and limitation. In carrying out its mission, it will comply with generally applicable Syrian laws and regulations and will not hamper the functioning of local civil administration. It will enjoy freedom of movement and communication and other facilities that are necessary for its mission. It will be mobile and provided with personal weapons of a defensive character and shall use such weapons only in self-defense. The number of the UNDOF shall be about 1,250, who will be selected by the Secretary-General of the United Nations in consultation with the parties from members of the United Nations who are not permanent members of the Security Council.

The UNDOF will be under the command of the United Nations, vested in the Secretary-General, under the authority of the Security Council.

The UNDOF shall carry out inspections under the agreement, and report thereon to the parties, on a regular basis, not less often than once every 15 days, and in addition, when requested by either party. It shall mark on the ground the respective lines shown on the map attached to the agreement.

Israel and Syria will support a resolution of the United Nations Security Council which will provide for the UNDOF contemplated by the agreement. The initial authorization will be for six months subject to renewal by further resolution of the Security Council.

APPENDIX B

THE 11 BRIGADES OF THE LEBANESE ARMY

The 1st Infantry Brigade is composed mostly of Shi'a Moslems in the Beqaa. It operates in close cooperation with Syrian forces in Beqaa and does not follow orders issued by the central Lebanese army command.

The 2nd Infantry Brigade is composed mostly of Sunni Moslems and is deployed in Tripoli. It is under the influence of Syrian forces.

The 3rd Infantry Brigade is composed mostly of Shiites and deployed in East Beirut.

The 5th Infantry Brigade is composed mostly of Christians and deployed in East Beirut.

The 6th Infantry Brigade is composed mostly of Shiites, most of whom are deserters from other Lebanese Army units. It is positioned in West Beirut and operates in coordination with the El Amal.

The 7th Infantry Brigade is composed mostly of Christians and positioned in North Lebanon. Soldiers in the brigade lean towards Lebanon's former president who is pro-Syrian.

The 8th Infantry Brigade is composed of a mixture of Moslem and Christian soldiers. Presently deployed in Souk-el-Arab, it is involved in daily exchanges of fire with the Druze forces.

The 9th Infantry Brigade is composed of a mixture of Moslem and Christian forces, and deployed in East Beirut.

The 10th "Airborne" Brigade is composed of a mixture of Moslem and Christian forces and deployed in East Beirut. Established as an elite airborne unit, its operational capability is in fact no greater than that of other Lebanese Army brigades.

The 11th Infantry Brigade is not yet in operation. It is made up of Druze soldiers and officers who deserted from the Lebanese Army and is situated at the Hamana camp (in the Metan). Some of these soldiers are presently siding with the Druze militias.

The 4th Mechanized Brigade was discharged in 1983, most of the soldiers having deserted. This brigade is not counted now.

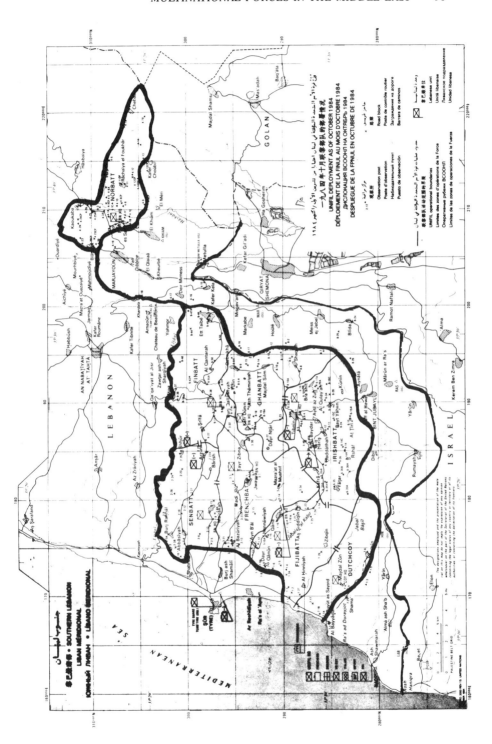

APPENDIX C

UNITED NATIONS
SECURITY COUNCIL

Report of the Secretary-General
Addendum

1.   Pursuant to paragraph D of the Agreement on Disengagement between Israeli and Syrian Forces (S/11302/Add.1, annex A), the Egyptian-Israeli Military Working Group of the Geneva Peace Conference under the aegis of the United Nations held six meetings in Geneva from 31 May to 5 June 1974. Military representatives of Syria joined the Working Group, and representatives of the Co-chairmen of the Conference also participated in the meetings.

2.   At the meeting held on 31 May, the military representatives of Israel and Syria signed the Agreement on Disengagement and a map attached to it. Following a brief intermission, the Military Working Group began work, in accordance with the Agreement, on the precise delineation of a detailed map and a plan for the implementation of the disengagement of forces.

3.   In the subsequent meetings, the Working Group reached full agreement on the following:

(a) A map showing different phases of disengagement;

(b) A disengagement plan and areas and a timetable;

(c) A statement read by Lt. General E. Siilasvuo, who presided over the meetings.

The map, to which the disengagement plan was attached was signed by the military representatives of Israel and Syria at the final meeting held on 5 June 1974. The agreed statement was also signed by General Siilasvuo at the same meeting, in conformity with an understanding between the parties.

4.   The plan of separation of forces involves the redeployment of Israeli forces from the area east of the 1967 cease-fire line. It also provides for Israeli redeployment from Quneitra and Rafid and the demilitarization of an area west of Quneitra still held by Israel.

5.   Prior to any Israeli redeployment, United Nations Disengagement Observer Force (UNDOF) will occupy, between 6 and 8 June, a buffer zone between the parties. The plan is to be implemented in the area of separation as specified in the Agreement. Separation of forces should be completed by 26 June. There is also provision for the return of Syrian civilian administration to the UNDOF area of separation.

6.   UNDOF will carry out an inspection of the redeployment of forces after the completion of each phase on dates fixed in the timetable attached to the plan of separation of forces and will report its findings forthwith to the parties. In order to determine that both parties have redeployed their forces in the limited forces areas, UNDOF will verify on 26 June 1974 that the limitation of forces agreed to by the parties is observed by the parties, and it will thereafter effect regular bi-weekly inspections of the 10-kilometer restricted forces areas.

7.   Agreement was also reached within the Working Group on the following points:

(a) Israel and Syria undertake to repatriate all prisoners-of-war still detained by them, not later than 6 June;

(b) Israel and Syria will co-operate with the International Committee of the Red Cross in carrying out its mandate, including the exchange of dead bodies, which is to be completed on 6 June 1974;

(c) Israel and Syria will make available all information and maps of minefields concerning their respective areas and the areas to be handed over by them.

APPENDIX D
UNITED NATIONS
SECURITY COUNCIL

LETTER DATED 15 MARCH 1978 FROM THE PERMANENT
REPRESENTATIVE OF LEBANON TO THE UNITED NATIONS
ADDRESSED TO THE PRESIDENT OF THE
SECURITY COUNCIL

On instructions from my Government, I have the honour to inform you as follows:

At midnight on 14/15 March, massive Israeli troops crossed into Lebanon along the Lebanese frontiers from several axes. The first was from Naqoura towards the village of Izziyah. The second axis was in the central sector, where Israeli troops reached the Tibnin heights. The third was on the heights of Rachayya al-Fukhar—Blatt—near Marjoyoun.

In addition to this naked aggression against Lebanese territory, Israeli patrol vessels penetrated Lebanese territorial waters along the coastline from Tyre to Sidon.

Furthermore, Israeli warplanes continue to fly in Lebanese air space and bombard the area.

An undetermined number of Lebanese citizens were killed, notably in Tyre, and enormous damage was caused to property. Consequently, large numbers of our people are leaving the south of Lebanon and going towards the north.

The Lebanese Government, while it vehemently deplores this aggression and protests strongly against it, wishes to make the following clarifications:

First, Lebanon had no connexion with the commando operation on the road between Haifa and Tel Aviv or with any other commando operation.

Secondly, it is a well known fact that Lebanon is not responsible for the presence of Palestinian bases in southern Lebanon in the present circumstances. The Lebanese

Government has exerted tremendous efforts with the Palestinians and the Arab States in order to keep matters under control. However, Israeli objections regarding the entry of the Arab Deterrent Forces to the south have prevented the accomplishment of Lebanon's desire to bring the border area under control.

Thirdly, the only solution to the problem lies in putting an end to Israeli aggression and in Israel's withdrawing its forces from Lebanon so that the Lebanese authorities can exercise their functions fully.

The Lebanese Government wishes to inform Your Excellency that it reserves its right to call an urgent meeting of the Security Council, and requests you kindly to have this letter circulated as a document of the Security Council.

(Signed)   Ghassan TUÉNI
Ambassador
Permanent Representative

APPENDIX E

UNITED NATIONS SECURITY COUNCIL

Report of the Secretary-General on the implementation
of Security Council Resolution 425 (1978)

1. The present report is submitted in pursuance of Security Council resolution 425 (1978) of 19 March 1978 in which the Council, among other things, decided to set up a United Nations Force in Lebanon under its authority and requested the Secretary-General to submit a report to it on the implementation of the resolution.

Terms of reference

2. The terms of reference of the United Nations Interim Force in Lebanon (UNIFIL) are:

(a) The Force will determine compliance with paragraph 2 of Security Council resolution 425 (1978).

(b) The Force will confirm the withdrawal of Israeli forces, restore international peace and security and assist the Government of Lebanon in ensuring the return of its effective authority in the area.

(c) The Force will establish and maintain itself in an area of operation to be defined in the light of paragraph 2 (b) above.

(d) The Force will use its best efforts to prevent the recurrence of fighting and to ensure that its area of operation is not utilized for hostile activities of any kind.

(e) In the fulfillment of this task, the Force will have the cooperation of the Military Observers of UNTSO, who will continue to function on the Armistice Demarcation Line after the termination of the mandate of UNIFIL.

General considerations

3.   Three essential conditions must be met for the Force to be effective. Firstly, it must have at all times the full confidence and backing of the Security Council. Secondly, it must operate with the full co-operation of all the parties concerned. Thirdly, it must be able to function as an integrated and efficient military unit.

4.   Although the general context of UNIFIL is not comparable with that of UNEF and UNDOF, the guidelines for these operations, having proved satisfactory, are deemed suitable for practical application to the new Force. These guidelines are, *mutatis mutandis,* as follows:

(a) The Force will be under the command of the United Nations, vested in the Secretary-General, under the authority of the Security Council. The command in the field will be exercised by a Force Commander appointed by the Secretary-General with the consent of the Security Council. The Commander will be responsible to the Secretary-General. The Secretary-General shall keep the Security Council fully informed of developments relating to the functioning of the Force. All matters which may affect the nature or the continued effective functioning of the Force will be referred to the Council for its decision.

(b) The Force must enjoy the freedom of movement and communication and other facilities that are necessary for the performance of its tasks. The Force and its personnel should be granted all relevant privileges and immunities provided for by the Convention on the Privileges and Immunities of the United Nations.

(c) The Force will be composed of a number of contingents to be provided by selected countries, upon the request of the Secretary-General. The contingents will be selected in consultation with the Security Council and with the parties concerned, bearing in mind the accepted principle of equitable geographic representation.

(d) The Force will be provided with weapons of a defensive character. It shall not use force except in self-defence. Self-defence would include resistance to attempts by forceful means to prevent it from discharging its duties under the mandate of the Security Council. The Force will proceed on the assumption that the parties to the conflict will take all the necessary steps for compliance with the decisions of the Security Council.

(e) In performing its functions, the Force will act with complete impartiality.

(f) The supporting personnel of the Force will be provided as a rule by the Secretary-General from among existing United Nations staff. Those personnel will, of course, follow the rules and regulations of the United Nations Secretariat.

5.   UNIFIL, like any other United Nations Peace-keeping Operation, cannot and must not take on responsibilities which fall under the Government of the country in which it is operating. These responsibilities must be exercised by the competent Lebanese authorities. It is assumed that the Lebanese Government will take the necessary measures to co-operate with UNIFIL in this regard. It should be recalled that UNIFIL will have to operate in an area which is quite densely inhabited.

6.   I envisage the responsibility of UNIFIL as a two-stage operation. In the first stage the Force will confirm the withdrawal of Israeli forces from Lebanese territory to the international border. Once this is achieved, it will establish and maintain an area of operation as defined. In this connexion it will supervise the cessation of hostilities, ensure the peaceful character of the area of operation, control movement and take all measures deemed necessary to assure the effective restoration of Lebanese sovereignty.

7.   The Force is being established on the assumption that it represents an interim measure until the Government of Lebanon assumes its full responsibilities in southern Lebanon. The termination of the mandate of UNIFIL by the Security Council will not affect the continued functioning of ILMAC as set out in the appropriate Security Council decision (S/10611).

8.   With the view to facilitating the task of UNIFIL, particularly as it concerns procedures for the expeditious withdrawal of Israeli forces and related matters, it may be necessary to work out arrangements with Israel and Lebanon as a preliminary measure for the implementation of the Security Council resolution. It is assumed that both parties will give their full co-operation to UNIFIL in this regard.

Proposed plan of action

9.   If the Security Council is in agreement with the principles and conditions outlined above, I intend to take the following steps:

(a) I shall instruct Lt. General Ensio Siilasvuo, Chief Co-ordinator of United Nations Peace-keeping Missions in the Middle East, to contact immediately the Governments of Israel and Lebanon and initiate meetings with their representatives for the purpose of reaching agreement on the modalities of the withdrawal of Israeli forces and the establishment of a United Nations area of operation. This should not delay in any way the establishment of the Force.

(b) Pending the appointment of a Force Commander, I propose to appoint Major-General E. A. Erskine, the Chief of Staff of UNTSO, as Interim Commander. Pending the arrival of the first contingents of the Force he will perform his tasks with the assistance of a selected number of UNTSO military observers. At the same time urgent measures will be taken to secure and arrange for the early arrival in the area of contingents of the Force.

(c) In order that the Force may fulfill its responsibilities, it is considered, as a preliminary estimate, that it must have at least five battalions each of about 600 all ranks, in addition to the necessary logistics units. This means a total strength in the order of 4,000.

(d) Bearing in mind the principles set out in paragraph 4 (c) above, I am making preliminary inquiries as to the availability of contingents from suitable countries.

(e) In view of the difficulty in obtaining logistics contingents and of the necessity for economy, it would be my intention to examine the possibility of building on the existing logistics arrangements. If this should not prove possible, it will be necessary to seek other suitable arrangements.

(f) It is proposed also that an appropriate number of observers of UNTSO be assigned to assist UNIFIL in the fulfillment of its task in the same way as for UNEF.

(g) It is suggested that the Force would initially be stationed in the area for a period of six months.

Estimated cost and method of financing

10.   At the present time there are many unknown factors. The best possible preliminary estimate based upon current experience and rates with respect to other peace-keeping forces of comparable size, is approximately $68 million for a Force of 4,000 all ranks, for a period

of six months. This figure is made up of initial setting-up costs (excluding the cost of initial airlift) of $29 million and ongoing costs for the six month period of $39 million.

11. The costs of the Force shall be considered as expenses of the Organization to be borne by the Members in accordance with Article 17, paragraph 2, of the Charter.

APPENDIX F

UNITED NATIONS SECURITY COUNCIL

RESOLUTION 425 (1978)
Adopted by the Security Council at its 2074th meeting
on 19 March 1978

The Security Council,

Taking note of the letters of the Permanent Representative of Lebanon (S/12600 and S/12606) and the Permanent Representative of Israel (S/12607),

Having heard the statements of the Permanent Representatives of Lebanon and Israel,

Gravely concerned at the deterioration of the situation in the Middle East, and its consequences to the maintenance of international peace,

Convinced that the present situation impedes the achievement of a just peace in the Middle East,

1. Calls for strict respect for the territorial integrity, sovereignty and political independence of Lebanon within its internationally recognized boundaries;

2. Calls upon Israel immediately to cease its military action against Lebanese territorial integrity and withdraw forthwith its forces from all Lebanese territory;

3. Decides, in the light of the request of the Government of Lebanon, to establish immediately under its authority a United Nations interim force for southern Lebanon for the purpose of confirming the withdrawal of Israeli forces, restoring international peace and security and assisting the Government of Lebanon in ensuring the

return of its effective authority in the area, the force to be composed of personnel drawn from States Members of the United Nations;

4. <u>Requests</u> the Secretary-General to report to the Council within twenty-four hours on the implementation of this resolution.

THE US FORCES' INFLUENCE ON KOREAN SOCIETY

Colonel
Lee Suk Bok
Republic of
Korea Army

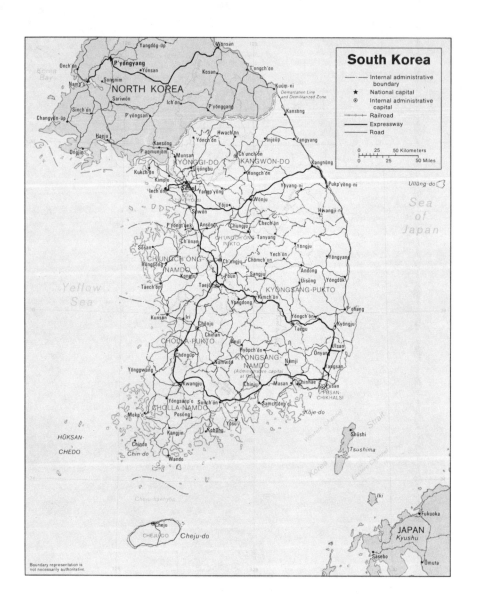

South Korea

— - — Internal administrative
          boundary
★ National capital
⊙ Internal administrative
     capital
┼┼┼┼ Railroad
━━━ Expressway
─── Road

0   25   50 Kilometers
├───┼───┼───┤
0   25   50 Miles

CONTENTS

TABLE

MAP

I

THE CLASH OF IDEAS

The presence of US forces in Korea brought about many changes in Korean society. While some were beneficial, other developments led to strain and to dissatisfaction with US influence.

Koreans found that the US forces had a tendency to use leverage and were inclined to intervene in Korea's internal affairs. Furthermore, the Americans often forced their will on the Korean government while Korea was fighting against communism. Vietnam's collapse demonstrated that such behavior can be dangerous and that the overbearing attitude of the US forces can sometimes help the enemy, instead of supporting a friend and ally.

Koreans understand the US desire to promote democracy in newly independent or developing countries, but they doubt that democratization is always well adapted to particular circumstances. Unless a country is leaning towards communism the United States should refrain from forcing her idealism on a foreign nation. US advice would be even more welcome in the developing nations if more forebearance were practiced.

The Sudden Introduction of US Mores

The South Korean people did not know how to deal with the sudden innovations introduced by the Americans. Suddenly, Koreans were confronted with American cowboy movies, broadcasting, Christian churches, pop music, relief material, US books and magazines—and the effect was overwhelming. In addition to the facets of American culture assimilated through the attitude of soldiers, movies, and music, a considerable shock resulted from the collapse of traditional Korean morale.

When the US forces disembarked in 1945 a new, democratic political system and a new social system were imposed on Korea.

Korea's dynasty lasted for 5,000 years, but the country had never known a cultural period under foreign influence that prepared Koreans to adapt to the sudden changes brought from the West. The systems introduced by the Americans were undeniably reasonable. They were an improvement on both the Korean dynastic government and Japanese colonial rule. Consequently, democracy was welcomed with enthusiasm by all South Koreans. However, Koreans were fundamentally unready to adopt democracy, and the sudden change resulted in chaos. US humanitarianism was good beyond compare in Korean eyes when they contrasted it with the past, and yet it accelerated the chaos in the country.

The American Forces in Korea Network (AFKN)

The American Forces in Korea Network was and is one of the foremost channels for the dissemination of American culture in Korea. The American Forces in Korea Network (AFKN) began broadcasting to the American soldiers in the front lines on 4 October 1950, immediately after the recapture of Seoul. From September 1951 it expanded and broadcast in nine languages for the sixteen countries participating in the Korean War. In 1957 six local stations were established and eight relay stations were also installed to ensure better reception throughout the Republic of Korea (ROK). In September 1957 the AFKN started to transmit TV and in July 1977 it introduced color screens.[1]

Korean society was greatly affected by the AFKN's introduction of various features of US society. The traditional Korean culture had already been severed and partially destroyed by the Japanese colonial policy. Therefore, when the AFKN was established Korean culture and society were already in a very vulnerable state. The AFKN played an important role in introducing new attitudes, hobbies, and a new way of expressing one's feelings, in addition to helping students learn English. While it was good to learn to understand the United States, the new enthusiasms had an adverse effect on the rehabilitation of Korean culture. The mixture of old and new cultures was frequently indigestible. It widened the gap between young and old. Today the question must be asked: Is there any way to reduce the influence of this broadcasting on Korean society?

The Dollar Economy

In addition to its impact on traditional life and mores, the US Forces in Korea (USFK) introduced a new, western-style economy. Three principal factors affected the Korean economy after the USFK arrived: Korean employment in US agencies, post exchanges, and local procurements.

Korean Employment in US Agencies Thirty-six thousand Korean employees worked in US units in December 1977; with the total US forces numbering 42,000 this was a considerable number. When the US 7th Infantry Division was withdrawn the US authorities fired 16,000 Korean employees.[2] How many Korean employees lost their jobs in 1954 when the second withdrawal was implemented can be estimated. In addition to these employees, large numbers of Koreans, in many different walks of life, relied on US soldiers near US bases for their livelihood. For example, the owners and employees of tailor shops, laundries, gift shops, bars, brothels, and so on were all dependent to some extent. In those days, the South Korean economy was considerably smaller than it is today and the Korean government did not have the capability to absorb unemployment caused by the US withdrawal. Sometimes, South Korean newspapers would report the suicide of a family that was caused by poverty, starvation, and unemployment.

Initially, US economic aid was concentrated too much on relief measures and consumer goods (80 percent) rather than on production and reconstruction (20 percent). The unemployment problem was solved gradually by the mid-1960s and reduced drastically by the mid-1970s, as the Korean economy developed from the time of the Military Revolution in 1961.

Post Exchanges The effect of the post exchange (PX) economy had both positive and negative aspects. PX goods were leaked into Korean society by US soldiers for the purpose of earning money for their entertainment expenses in the early years. Later on, blackmarketeers collaborated with PX employees and mass leaks occurred. These leaks from the PX were referred to as the "PX economy." When the PX economy prevailed about 60 percent of total sales was believed to flow into the Korean society. On the positive side, the PX economy provided the materials which were in short supply in the ROK and thus it prevented post-war inflation.

On the negative side, the PX economy created confusion in the ordered Korean environment, fostering a taste for unbridled consumption and hindering domestic industries. It created various social abuses such as conspicuous consumption, foreign tastes, and the faking of foreign goods. Since the US-ROK Status of Forces Agreement was concluded in 1966, ROK and US joint efforts have combined to prevent the leaking of PX goods. In 1975 the US authorities restricted use of the PX to dependents, and agreed to punish those who were selling and buying merchandise leaked from the PX. Nevertheless, the PX economy still hides underground, even though it has diminished. This fact shows clearly how difficult it is to cure bad habits once they have become rooted in society.

Local Procurement Local procurement of the necessary material for the USFK was begun in 1955 when the official foreign exchange rate was established. Previously, all materials had come either from the continental United States or Japan. However, at that time South Koreans were not properly prepared to sell goods and services to the USFK. The people of the US procurement agency even had to teach Koreans how to cultivate sanitary vegetables, how to follow US procurement specifications, etc. The great construction companies, such as Hyundai and Daelim, which now compete with advanced countries in the international market, got their start in construction work for the USFK. The ROK government realized the significance of military supplies for the United States and from 1962 onward supported the growing firms positively. The earnings of the military supply and service contract firms were almost equivalent to the total amount of the exports of all of Korea in the early years.

COMPARISON OF EXPORTS AND MILITARY SUPPLY EARNINGS
(millions of dollars)

Years	GNP	Export	Military Supply
1961	2,103	40.9	38.4
1962	2,315	54.8	34.0
1977	37,429	10,046.5	140.0

Source: Bank of Korea

The military supply contracts for the USFK stimulated the development of the South Korean economy in many ways. For example, because US Forces emphasized the sanitary cultivation and treatment of food supplies, this attitude encouraged improvements in the Korean diet. Furthermore, a lot of military supply contracting firms became big companies and were able to create a tremendous number of jobs.

Later, when Secretary of Defense Caspar W. Weinberger and Minister of National Defense of the ROK Yoon, Sung Min agreed on the participation of the Korean defense industry in the maintenance program for USFK equipment at the 17th Annual Security Consultative Meeting in May 1985, they shared the view that the capabilities of the Korean defense industry were important defense resources, not only for Korea, but also for the Free World. In addition, such participation will further the growth of the Korean defense industry.

Besides the cultural and economic changes, there were changes in the pattern of personal relations after the coming of the Americans.

II

SOCIAL CONSEQUENCES OF THE WAR

Relations between men and women in Korea were formerly very discreet. Moreover, until the time of the US Forces' arrival, intimate relations between foreign men and Korean women were regarded as almost sinful. For example, in October 1945 there was a show to welcome the US Forces and when a female sang, the Korean audience ridiculed and blamed her for singing before foreigners. As a result, the show couldn't continue.[3] However, starvation and the death of husbands in the war ruined the old customs, traditions, and ethics. Many women sold their bodies to US soldiers in order to survive.

Mixed Blood

As a result of tragic wartime circumstances many Korean women became the prostitutes of the Western troops. Babies of mixed blood became a social problem. There are believed to be about twenty-five thousand children of mixed blood in Korea.[4] The cherished desire of these unhappy children is adoption by the American parent because in the United States they would not be discriminated against whereas they are treated with contempt by the homogeneous Korean society. Those of mixed blood do not even have to perform compulsory military service with other Korean youths. Fortunately, many children of mixed blood have been adopted, but quite a number are still living in Korea under the shadow of their heritage.

Transcultural Marriage

As the Korean society gradually became more open, marriages between American soldiers and Korean women increased. Many Korean women who married American soldiers were employees of US Forces' units who had chances to contact and understand American soldiers, and some were women who were having difficulty in finding a Korean husband. Women who were divorcees or widows faced this problem, unlike their contemporaries in the United States. Approximately sixty thousand marriages have taken place, with an annual rate of about three thousand marriages. Some of these marriages face problems as a result of the different cultural backgrounds and diverse behavioral patterns. In such cases neither the American nor the Korean partner might be wholly to blame for the breakdown of the marriage.

However, the majority of transcultural couples achieve a happy married life and often Korean wives invite their families from Korea to join them and ease their own stress in a new, strange world. This practice has caused considerable immigration, since quite often a Korean wife will bring three family members to the United States.

III

THE BEHAVIOR OF AMERICANS IN KOREA

In Korea the impact on an Asian society of US behavioral and political patterns has not always been happy. When, in September

1970, Dean K. Froehlich from the Human Resources Research Organization wrote a technical report for the chief of Research and Development, Department of the Army, with the title of "Military Advisors and Counterparts in Korea" (a study for personal traits and role behaviors), he concluded, "The Koreans want their advisors to display more often an interest in becoming knowledgeable about the country's language, history, economy, customs, and the feelings of the Korean people."[5]

South Korea has never said, "Yankee go home!" Korea's attitude to the Americans in Korea, however, demonstrates not only the Koreans' fondness for Americans but also the traditional Korean hospitality. The customs and values that derive from Confucianism make it incumbent on Korean hosts to deal with guests hospitably. Some Koreans blame Americans for behavior which is despised by Koreans. And, it is true, Americans do not often put themselves in Koreans' shoes. The ability to do this occasionally is very important in working effectively with Koreans, and it reduces the chances of friction or misunderstandings that arise between different cultures.

The two nations are likely to be working together for some time so this is an important consideration. After examining the historical role of the American forces in Korea we should consider future possibilities.

IV

THE FUTURE ROLE OF THE USFK

The Perspective for Withdrawal

With every change in US administration there are always some changes in the policy of the USFK. Koreans would like to see a firm and consistent US policy. Therefore, a long-range plan, agreed upon bilaterally, for the withdrawal or presence of the USFK should be devised now. Once such a plan was established, it need not be fixed forever but could be subject to discussion between the two countries whenever the situation changes.

The Question of Military Balance

The ROK's lack of self-defense capability has tempted North Korea to build up its own forces to menace South Korea. Furthermore, balance in the Korean peninsula, based on ROK-US combined combat power, hinders political negotiations because North Korea insists on the withdrawal of the USFK before any further negotiations can take place. If a move toward political negotiation were to occur now it would be only a temporary measure on the part of North Koreans, undertaken with the intention to deceive. But when the ROK achieves an adequate defense capability in the 1990s real political negotiations can be undertaken. It is worth discussing now whether certain changes might be made at that point which would improve the ROK's international stature and raise the ROK's standing in comparison with North Korea. For example, would it be possible to substitute a Korean general for a US general as chief of the armistice committee of the UN command? Would there be any problem in changing the commander position of the ground component under the Combined Forces Command from a US general to a Korean general? Whether it would be feasible to proceed without regard to North Korea's probable opposition, would be a question worth discussion at future SCM meetings.

If the ROK achieves a military capability equivalent to that of North Korea in the 2000s, the presence of US Forces in Korea, especially US ground forces, will not be necessary for the purpose of defense against North Korea's provocation. However, the United States may need to stay in Korea for other reasons. The broader US strategic goals, such as checking Soviet expansionism, will not change. In the event that US forces remain in South Korea the ROK will agree with the United States according to Article 4 of the Mutual Defense Treaty but it will not want to be hindered in conducting negotiations toward unification with North Korea.

The question of the continued existence of the UN Command and the Combined Forces Command in this time frame needs to be addressed. New arrangements in the relationship between Korea and the UN Command, CFC, and USFK will be likely in this period.

The Present Role of the USFK

The ROK, for its part, views the role of the USFK as:

(1) Maintaining the balance of power among the big powers in Northeast Asia in order to check Soviet expansionism in the Asia/Pacific region, and to protect Japan and cooperate in containing the Soviet and the PRC.

(2) Acting as a deterrent to war in the Korean peninsula, to prevent North Korean adventurism, and to suppress the ROK's possibly excessive military actions against North Korea.

(3) Protecting the US political and economic interests in the Northeast Asia region.

(4) Demonstrating symbolic determination to defend Asia and the Pacific area.

(5) Contributing to the security of Western interests by dispersing the Soviet military power in Northeast Asia and maintaining the security of Northeast Asia.

(6) Contributing to the development of the ROK by assisting the development of the ROK economy, by developing the military skill of the ROK Armed Forces, and by stabilizing South Korean society.

From the Korean point of view, the USFK presence has hitherto had effects that are partly adverse. The presence of the USFK has encouraged North Korea to constantly increase its military power and it tempts the North Koreans to resort to nuclear armament. Furthermore, the USFK slows the development of a self-reliant defense policy and military strategy in the ROK, and the USFK presence causes an unbalanced military power structure within the ROK armed forces. The Soviets and the PRC have been forced to support North Korea as a counterweight to the presence of the USFK; this presence is used as leverage to intervene in the ROK's internal affairs. Finally, the US forces cause some cultural friction in Korean society.

The Future Role

The role of the USFK is unlikely to change greatly in the future. Meanwhile, elements (2) and (6) from the present role might be reduced in intensity; steps could be taken to reduce some of the adverse effects that stem from a clash of cultures. In addition, the USFK should assist the ROK to play a bigger role in the defense of the Korean peninsula. Were the ROK's power equivalent to that of North Korea, the United States could concentrate its power on checking the Soviets. This, in the author's opinion, would be the fastest, best way to ease the tension in the Korean peninsula and to bring the North Koreans to the table of political negotiation.

V

PROS AND CONS

American interest in the ROK stems from the fact that Korea, one of the most strategic confluences in the world, has a special relevance to the global balance of power. It is the only place where the direct interests of four major world powers interact. Korea is geographically critical in the Far East, being a peninsula where the conflicting interests of several major powers have met for centuries. The peninsula has been thought of by the Japanese as "A dagger pointed at the heart of Japan." With equal logic it has been seen as "a hammer ready to strike at the head of China." For the Soviet Union, Korea commands the port of Vladivostok.

When the US forces disembarked in Korea in 1945, neither the US government nor the US military fully understood Korea's geostrategic value. No detailed, clear directions were provided to the members of the USFK charged with disarming the Japanese and preserving law and order until the Koreans themselves could take over. The planned political process was delayed as a result of a lack of information about the history and culture of Korea. There was repeated trial and error because the Americans had no grasp of the Korean way of thinking and failed to understand Koreans' bitter feelings against the Japanese.

From the beginning of the occupation to the first withdrawal of the US forces, the US military government, with no understanding of the peculiarities of Korea, focused its energies on implanting American ideas and democratic principles in Korea.

Chaos resulted, with a succession of demonstrations. Within two months of the arrival of the US military government there were as many as 250 groups formed by political parties and military factions. Such was the emergence of democracy in the new Korea under the control of the US military government. Later, the first withdrawal of the USFK, without sufficient strengthening of the ROK armed forces, brought about the Korean War.

Indisputably, the United States failed to understand, unlike the USSR and the PRC, the strategic importance of the Korean peninsula. Nonetheless, Koreans must always appreciate the efforts of the Military Advisory Group in Korea (KMAG), who devoted themselves, from 1945 to 1948, to activating and training the ROK Army soldiers and units. The creative activity of these men who activated the Constabulary, while the US government continued to postpone the decision on the recommendation for the 45,000-man Korean national defense force, merits high praise. USFK personnel provided direction in the field, whereas the higher echelons of the US government simply procrastinated.

The US forces in Korea had the greatest impact on Korean society, greater than any other foreign presence in her history. Even the 35 years of Japanese colonial rule had less influence on Korea than did the US forces, who were never autocratic. The USFK brought a new wind, "American style." The new wind created a whirlwind of democracy in the political and social systems. The modernizing trends clashed with intolerant customs and primitive industries. South Korea's whole culture and life style were suddenly and drastically westernized.

Despite the clash of cultures, it must be acknowledged that the presence of US Forces in Korea has deterred another war in the Korean peninsula and has clearly enhanced stability in Northeast Asia. The danger in the Korean peninsula is not simply that in the near future North Korea might launch a massive military attack against the South, either on its own initiative or at outside instigation; the real danger is that the Soviet Union will never be content with the preservation of rough equilibrium between the communist sphere of influence and the Western sphere. The Korean peninsula provides a decisive flank that obstructs the Soviet Union's designs to encircle the PRC and Japan. As the ROK Armed Forces progressively gain the capability for self-defense against North Korea, US forces in Korea can begin to concentrate their efforts on checking the Soviet expansionism—starting in the 1990s, with any luck.

Both Koreans and the US military have learned a great deal about each other's worlds through their close proximity in Korea and, with hindsight, it should be possible to avoid the mistakes which earlier marred the two countries' mutual involvement.

ENDNOTES

1. Seoul Newspaper, pp. 427–429.
2. Seoul Newspaper, pp. 416–418.
3. Seoul Newspaper, p. 430.
4. Seoul Newspaper, pp. 454–456.
5. Dean K. Froehlich, *Military Advisors and Counterparts in Korea: A Study of Personal Traits and Role Behaviors.* (Washington, D.C.: Human Resources Research Organization, for the Department of the Army, Chief of Research and Development, September 1970) p. vii.

THE SOVIET INVASION
OF AFGHANISTAN AND ITS
IMPACT ON PAKISTAN

Brigadier
Zia Ullah Khan
Pakistan Army

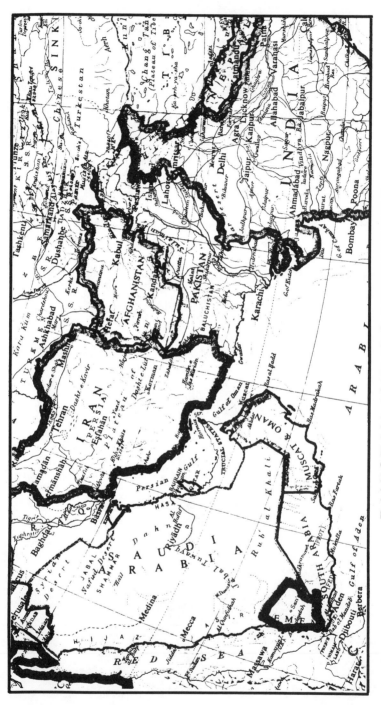

PAKISTAN AND ITS NEIGHBORS

CONTENTS

MAPS

Southwest Asia

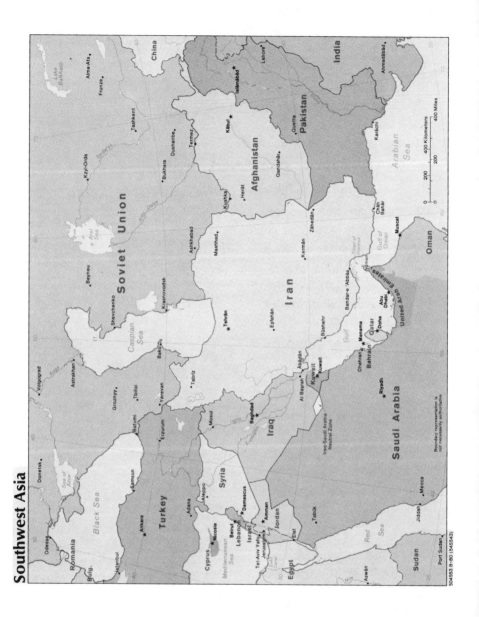

I

PAKISTAN AS A FRONT-LINE STATE

The Soviet thrust southwards into Afghanistan, long the buffer between contending powers, shattered the shield that prevented confrontation and brought an era of power struggle to the region. There may conceivably be a myriad of explanations or excuses for the Soviet advance into Afghanistan; however, in extending their hold and influence in that country the Soviets have undoubtedly gained an enormous geostrategic advantage. They have, in fact, furthered their historic objective of reaching warm water and their ultimate goal of controlling a wider area.

Pakistan, which previously had the advantage of Afghanistan as a buffer with the Soviets, is today a front-line state directly facing the Soviets. Pakistan remains a serious and probably the last impediment in the way of Soviet ambitions. When the changed environment that resulted from the Soviet maneuver began to be apparent, Pakistan was itself in serious difficulties with an irreconcilable neighbor, India, on its eastern border and an unstable internal situation. The occupation of Afghanistan by the Soviets, with all its implications for the future, added another dimension to the serious security situation which already existed in Pakistan.

The Soviet invasion of Afghanistan has confronted Pakistan with some very great challenges. At the same time, Pakistan stands out as the last post in the free world's aspiration to contain the Soviets in this complex situation. Pakistan, caught in a quagmire, is engaged in an all-out struggle to safeguard its freedom and live up to the hope of like-minded freedom-loving people of the world. The redeeming features in the midst of many difficulties are Pakistan's leadership and its people who have continued to show great resolve and determination in meeting seemingly insuperable problems.

The paper analyzes the impact of the Soviet invasion of Afghanistan on Pakistan and the Soviets' future aims and objectives

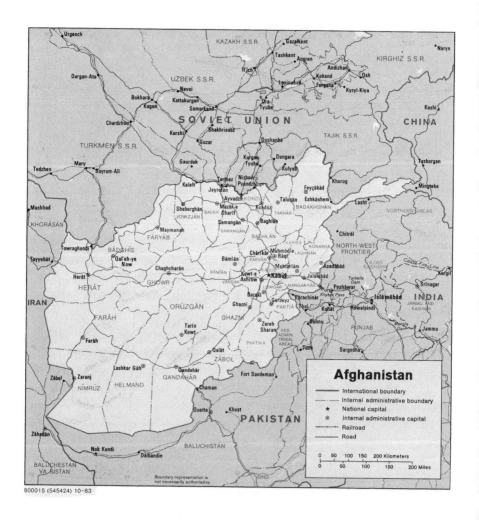

Afghanistan

— International boundary
--- Internal administrative boundary
★ National capital
◎ Internal administrative capital
— Railroad
— Road

0   50   100 150   200 Kilometers
0   50   100   150   200 Miles

Boundary representation is
not necessarily authoritative.

800015 (545424) 10-83

126

in the region and their probable effect. Touching only very briefly on the ways and means of checking Soviet ambitions, I have left it to the readers of this paper to draw their own conclusions as regards finding the best way of meeting the challenges of the future.

II

BACKGROUND TO THE SOVIET INVASION OF AFGHANISTAN

The conquests to which the people of Afghanistan have been subjected over the centuries are legion, from that of Alexander the Great in 331 B.C. to the Soviet invasion of December 1979. As many as twenty-five dynasties have ruled this country—from the Aehaemenian of Cyrus and Darius in the sixth century B.C. to the Muhammedzais, the last of whom, General Mohammad Daud Khan, put an end to monarchy in Afghanistan with a family feud that toppled his cousin, King Zahir Shah, in July 1973. This coup was the catalyst for the beginning of the end of Afghanistan's independent status.

The coup of 1973 when King Zahir Shah was dethroned by his cousin Daud came as a surprise, since King Zahir Shah was himself on good terms with the Soviets. However, it is certain that the Afghan monarchy could not have been overthrown without the knowledge, if not the connivance of the Soviet Union. The coup may have been prompted primarily by Daud's ambition and his assurances of a future posture far more pro-Soviet than the king's, but the coup also involved the unfolding of a big game—the successor to the Great Game played in the nineteenth century between the Russians and the British—being played by the Soviets. Soon after taking over, Daud abandoned Afghanistan's traditional neutrality. The Soviets assumed that he would support their policies in international affairs. However, events turned out differently than expected and, in 1978, Daud veered in the other direction and displayed a pro-Western tilt, providing the Soviets with a chance to remove him in a bloody coup in which his entire family was killed. As happened in the past whenever

Afghanistan abandoned or compromised its neutrality in international relations, Afghanistan now experienced dire consequences.

Following Daud, a Marxist regime was installed in April 1978, much against the wishes of the people of Afghanistan. Despite Soviet political, economic, and military support, the regime got into serious problems at the very outset. In fact, the authority of this regime was contested in almost all the 28 provinces of Afghanistan. The Afghan populace, trained over centuries in individualism, began to make life difficult for the government through a variety of responses, ranging from absenteeism from work to individual acts of terrorism against Marxist officials. At the same time, large numbers of Afghans began migrating to Pakistan. The change from Nur Mohammad Taraki's regime to that of Hafizullah Amin brought no relief. Finally, in December 1979, the Soviets, finding the situation getting out of control, moved into Afghanistan with troops and installed their most trusted protégé, Karmal. Since then the Soviets have been fully involved in administration as well as in military actions to regain control of the situation. Afghanistan is not the first country to be occupied by Russia. In fact, it is the seventh Islamic state to be so occupied since the outbreak of the communist revolution in 1917. Three of these states are situated in Central Asia (Kazakhastan, Turkmenistan, and Uzbekistan), two on the frontier with China (Tadjikistan and Kirghiztan), and one, Azerbaijan, is situated on the frontiers of Turkey.

The occupation of Afghanistan by the USSR rocked the entire region which immediately understood the possible ramifications of the Soviet move. At the same time, the geostrategic situation of Pakistan was transformed. Instead of being buffered by the complex terrain of Afghanistan, which had so long separated Russian and then Soviet territory from the sub-continent, Pakistan now faced Soviet troops virtually anywhere along the 1,300-mile frontier. The shadow of Soviet power now hung over the whole of the sub-continent as never before.

Pakistan, a nation of 85 million people, shares its borders with India in the east, with the People's Republic of China (PRC) in the north, and with Afghanistan—now, for all practical purposes, the USSR—in the west. To the west and southwest lies Iran which, with India, flanks Pakistan. Iran, like Pakistan, has an Indian Ocean

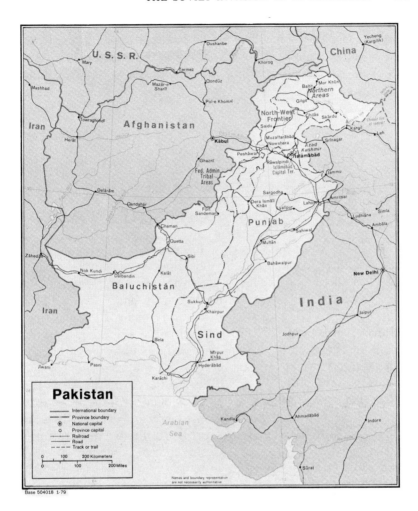

coastline. Geography has thus placed Pakistan in one of the world's most sensitive regions; it lies physically at the mouth of the Persian Gulf. Pakistan can be described as a last barrier to Soviet ambitions in Southwest Asia.

III

SOVIET AIMS AND OBJECTIVES

In 1717, Peter the Great, in formulating his expansionist designs, was the first to focus his attention on Southwest Asia. He stressed the necessity of gaining access to this region—entering into the so-called "Warm Waters" of the Indian Ocean. Today, the Soviet Union is the world's largest state, covering one-sixth of the earth's land mass, an area of 8,647,250 square miles, stretching from Eastern Europe across North Asia to the Pacific, with over two-thirds of Soviet territory, conquered during the last four centuries, lying in Asia. To the South, the USSR is bordered by Turkey, Iran, Afghanistan, China, Mongolia, and North Korea. Of the six countries lying on the southern border, only three—Turkey, Iran, and China—are outside the Soviet orbit. With the occupation of Afghanistan Pakistan, too, has become a state bordering the Soviet Union and it is today struggling to remain outside the sphere of Soviet influence. These geopolitical realities will have to be kept in view while pondering the future course of events.

Czarist Russia moved forward in the direction of Afghanistan after occupying Turkestan (Turkemanis) and the Kharrakes of Central Asia. The Russians came to a temporary halt for reasons of consolidation and as a result of British military and political measures to keep Afghanistan as a buffer to protect their Indian colony against Russian aggression. The Soviets never gave up their struggle for expansion in this direction and, finding the situation in their favor struck, in December 1979, to occupy Afghanistan, thereby seizing a strategic zone only 400 miles away from the "Warm Waters"—with Pakistan situated in between the Soviets and their objective. There were probably three major reasons for the Soviet takeover.

First, the Soviets feared that the revolution in Iran and the resurgence of Islam in the region would spread to Afghanistan and from Afghanistan to the Soviet Central Asian Republics. The Central Asian states of Russian Azerbaijan, Tadjikistan, and Uzbekistan have a substantial Muslim population. Historically, they have been the center of Muslim art and culture for centuries. Even today, the ruins of some of the buildings in Samarkand, Khiva, Tashkent, and Bokhara are a testimony to the advanced skills of the people who inhabited these regions. After their assimilation into Czarist Russia during the eighteenth and nineteenth centuries, the people of these areas remained linked to their Muslim heritage and traditions. Nearly two centuries of occupation have not succeeded in breaking this bond. Hence the Soviet dilemma after the revolution in Iran.

The second reason for the Soviet actions, partially related to the first, suggests that Daud, who had overthrown King Zahir Shah of Afghanistan in 1973, was becoming lukewarm towards the Soviets and was strengthening his ties with the Shah of Iran. The Soviets, therefore, engineered his departure and the induction of a communist regime. The new regime rapidly established both its minority character and its unacceptability to the Afghan people but the Soviets, once having committed their support, could not withdraw that support without loss of prestige. The Soviet invasion of 1979 was thus an exercise in the maintenance of Soviet prestige after the initial miscalculation of instigating and supporting the communist coup of April 1978.

The third and most commonly quoted reason for the occupation of Afghanistan is that it was a further step in the fulfillment of the last testament of Peter the Great in which he had instructed his successor to continuously probe southwards for warm water ports. If one looks at a map, it is obvious that by occupying Afghanistan, the Soviet Union has created a wedge which gives it multiple strategic advantages in relation to Southwest Asia and South Asia. Finally, apologists for the Soviet Union and those concerned about a Grand Soviet Strategic Design agree that by moving into Afghanistan the Soviet Union has positioned itself superbly for taking advantage of any opportunities that may arise in the future. Whichever of the above explanations we accept, the inescapable conclusion is that a buffer state, which hitherto had put a distance between a superpower and South Asia and, to a lesser extent, between a superpower and the littoral states of the Persian Gulf, is being made into a satellite, ruled by a puppet regime. Soviet power today is poised at the entrance of South Asia and this extension of Soviet power affects not only South Asia but also the Persian Gulf with overall serious ramifications.

All reports coming out of Afghanistan confirm Soviet attempts to consolidate this strategic advantage. For instance, the railway line from Russia has been extended into Afghanistan across the River Oxus, permanent barracks, warehouses, and new bridges have been constructed and highways are being improved. A huge Soviet airbase at Shindand is under construction and is fortified by minefields. The minefields are to protect the airbase which is believed to include missiles, possibly missiles with strategic ranges. The Soviet military strength has increased to 1,250,000 troops.

Pakistan today is one of the two countries lying between Soviet Russia and the fulfillment of its dream of reaching the warm waters. Should the Soviets succeed in their design or manage to slice through Pakistan, with or without the consent of the people of Pakistan, the consequences for the free world would be grave. The Soviets would acquire the capability to physically block the Strait of Hormuz and thereby exercise control over the oil supply for Western Europe and Japan. In such an eventuality the implications for the free world, including Japan, would be very serious. In the military field, the words of Leon Trotsky in 1919 are apropos, "The road to Paris and London lies through the towns of Afghanistan, the Punjab, and Bengal." The Soviet gunships and airborne divisions are well poised in Afghanistan for sallying out towards the Indian Ocean. No doubt such an occurrence is a bit futuristic and contingent upon the stabilization of the situation in Afghanistan, but unless Soviet difficulties in Afghanistan today are accentuated and peripheral states are sufficiently strengthened this contingency will materialize sooner rather than later.

A firm foothold anywhere on the coast of the Indian Ocean would certainly enable the USSR to turn the ocean into its private lake. The Soviets would thus have made a breakthrough of strategic dimensions. Moscow having, by one stroke, completed the encirclement of one of its adversaries, China, would find easy access to the Middle East. It would then be in a position to threaten NATO's southern flank beyond its capacity to sustain, and, last but not least, the USSR would also be in a position to overcome its major vulnerability (one that constantly nags its planners) to nuclear attack from the US submarines now moving about freely in the Indian Ocean. In addition, this extended Soviet presence would enable them to generate enormous political, economic, social, and cultural influence in major parts of the world and would bring them close to their ultimate objective of world domination as laid down in their doctrines.

IV

THE IMPACT ON PAKISTAN

Pakistan was already grappling with large numbers of serious security, political, economic, and social problems when the tremor of the Soviet's move shook the country, accentuating the existing dilemma. In the existing security environment the country could hardly afford to provide minimum defensive capability against the eastern border with India, yet it was suddenly caught up with a two-front war scenario with unfriendly India on its eastern border and the hostile Soviets poised on the western border. A rather unstable situation on the home front further complicates the security situation. In these circumstances the government must show caution to prevent irredentist elements, sponsored by outside powers, from achieving their ends.

Economically, Pakistan is facing far greater challenges than ever before. With limited resources at its disposal, it is required to meet the changing security requirements and, at the same time, keep its teeming millions satisfied in order to prevent unhealthy influences creeping into society. The influx of large numbers of Afghan refugees has further complicated the existing economic problems besides having a considerable impact on the political, social, law and order, and security spheres. The good news is that whereas the nation faces problems of a serious nature resulting from the Soviets' initiative, Pakistan has certainly manifested great powers of will and dignity to meet the challenges of the eighties.

The Threat From India

Since independence, Pakistan has fought three wars with India and experienced countless border clashes. In the last Indo-Pak War of

1971, India struck when Pakistan was deep in the mire of an extensive internal insurrection and the result was the dismemberment of Pakistan. The deep-seated distrust and hostility that exists between Muslims and Hindus stems from two very different ideologies. Partition was inevitable for the health of these two nations. However, the incidents of communal carnage that preceded and climaxed the partition of the British Indian territories, and the resolve of Hindu leaders to undo partition by any means, further hardened the attitudes of both nations vis-à-vis each other. Moreover, Mountbatten's desperate surgery had left many grave issues unsettled.

Among these the Jammu and Kashmir question survived as the main cause as well as the symbol of India-Pakistan animosity and intransigence. It was this sense of insecurity which compelled Pakistan to a search for allies, a search later manifested in the form of an alignment policy. Pakistan's participation in the Western defense alliance systems angered and frustrated the Indians and they began consolidating their military hold over Kashmir. Instead of resolving the dispute, India accelerated the erosion of Kashmir's special status and gradually integrated Kashmir into the Indian Union. Pakistan's repeated protests were ignored by India. Consequently, the Kashmir dispute became a more serious source of friction and antagonism between the two neighbors and shaped into a major facet of the Indian threat to the security of Pakistan.

While the East Pakistan crisis demonstrated India's unabashed willingness to intervene militarily in Pakistan's internal affairs, the separation of East Pakistan, in fact, improved Pakistan's security situation. As a result of the 1971 debacle, Pakistan was reduced in size and population but not reduced significantly in military strength. In spite of the improved strategic position vis-à-vis India, Pakistan's security dilemma remained acute. Not only is India vastly superior in numbers, it has also a well-developed arms industry. Pakistan does not really have any arms industry worth mentioning and is, therefore, heavily dependent upon outside suppliers for military hardware. Ever since Pakistan's withdrawal from SEATO and CENTO and the imposition of a US arms embargo, the procurement of arms has been difficult. Assistance from the Peoples Republic of China helped the rebuilding of Pakistan's military strength in the post-1972 period. However, sophisticated modern arms are extremely costly and tend to age quickly and, therefore, have to be replaced early. Pakistan's economy is not strong enough to sustain a regular inflow of modern arms. Despite the recently renewed links with the United States, the intake of modern arms is not commensurate with the security challenges that Pakistan faces today. On the other hand, India's well-equipped armed forces are more than three times larger than Pakistan's, and are backed up by an indigenous arms industry and dependable external suppliers.

Another dimension of Pakistan's perception of the Indian threat that needs to be mentioned here is India's acquisition of a nuclear capability. The Indian nuclear explosion generated a new wave of fear of possible future nuclear blackmail among Pakistanis who were conscious of the past Indian attitude vis-à-vis Pakistan. Pakistan felt that not only had India always enjoyed a numerical superiority in armed personnel and conventional arms but that by going nuclear it had acquired qualitative technological superiority. This meant that henceforth Pakistan would have to live under the shadow of a hostile and powerful nuclear neighbor. This is indeed a bitter pill to swallow.

Given the existing disparities in size, population, resources, technological development, and military capabilities, as well as India's non-conciliatory attitude, the threat from India, measured by any yardstick, continues to be real and serious. Current Indian reactions to Pakistan's efforts towards modernization of its forces, the muted Indian response to the Soviet invasion of Afghanistan, and the discouraging attitude towards Pakistani peace moves in the area are further cause for concern. At the same time, India's leaning on the USSR as a partner in the Soviet's friendship treaty, her position as the recipient of generous military aid from the USSR, and her continued hostility towards Pakistan combine to confront Pakistan with an extremely dangerous potential scenario of a two-front war.

The Threat from Afghanistan

The crossing of the Oxus River by the Red Army in December 1979 lends credence to the interpretation that the Soviets are on the path of expansionism and, by implication or by design, Afghanistan may turn out to be only the first target in this direction. However, five years of occupation have not yet brought control of the situation in Afghanistan. Were the Soviets prepared for this protracted warfare, or have they been surprised? Whatever may be the case, the Soviet's past record shows clearly that they do not give up. In any case, their leap forward places the Soviets strategically well poised for achieving the ancient Czarist desire.

By dominating Iran and Pakistan the USSR could eventually gain control of this region and improve its strategic posture. It is unlikely that the Soviets will be able to achieve their maximum objective by direct action without some serious political and military responses from the other powers. Even while keeping the option of direct action open, they are likely to employ indirect means—use of proxies, covert activities, and support of separatists to achieve their ends. There is historical proof that the Soviets are averse to unfriendly, uncooperative small neighbors. Sooner or later such nations are removed from the Soviets' path. Pakistan stands out as an obstacle to Soviet ambitions in Southwest Asia, and must be humbled, primarily in the interests of the stability of the communist regime in Afghanistan.

Beside the option of direct military action there are three other possible options which the Soviets may exercise against Pakistan in the existing situation. The first option arises from the presence of large numbers of Afghan refugees on Pakistani soil and the on-going Afghan Resistance War which may draw Pakistan into the Afghan cauldron, albeit unwillingly. Assuming that the civil war persists and the Soviet casualty rate increases significantly, there may come a point where the Soviets might seriously contemplate hot-pursuit and sanctuary busting operations. Once this happens, Pakistan would come under greater threat of cross-border operations and other intimidation tactics. From the Soviet perspective, this option would be entirely defensive. What the Soviets appear to have failed to recognize is the fact that the massive refugee influx has presented the government of Pakistan with an irreconcilable dilemma. If it organizes help and provides the bare minimum of facilities for these refugees on humanitarian grounds, then Moscow will accuse Pakistan of aiding, abetting, and encouraging what it terms "counter-revolutionary elements." If Pakistan does not look after them there is the danger that the refugee camps will become hotbeds of insurgency and may even alienate the local population of the area because of their deep feeling of Muslim brotherhood.

Another Soviet option would be to exploit the suspicion and antagonism existing between India and Pakistan as a means of increasing pressure on Pakistan. While direct military action by the Soviet Union against Pakistan is likely to invoke reaction, a clash with India would be much less likely to do so, for the obvious reason. However, a complementary maneuver by the Soviets on the Western borders to tie down or divert some of Pakistan's forces could be undertaken to ensure the success of the Indian armed forces in achieving their military objective. This is a two-front war scenario with extremely serious implications for Pakistan. India, meanwhile, appears to be quite predisposed to play the Soviets' game for its own reasons.

The third option would be the creation of an independent Baluch state with the exploitation of dissident elements already existing in Baluchistan. This would be a good bargain since it would give the Soviets the right to move about freely to approach the only port, Gwadar, which lies directly across from the Strait of Hormuz, an economic choke point of world importance at the entrance of Persian Gulf. This option might be combined with the second option and exercised at a point in time when Pakistan's armed forces were thoroughly committed elsewhere.

V

PAKISTAN'S INTERNAL SITUATION

Some of the major vulnerabilities of Pakistan emanate from internal problems. Among the host of domestic problems, three appear to be the important ones: the continued search for a viable political system; the unstable performance in the economic field; and the want of national cohesion.

The Political System

The continued failure of Pakistanis to maintain a viable political system has not only consistently impeded the development of nation-building institutions, but has also generated a number of complex problems. In this sphere excessive political experimentation at various periods of Pakistan's history by different leaders has not yet provided a solution for the political problems of Pakistan. Three times the Army has had to take over the country in order to rid the nation of the abuse of power and the malpractices of politicians and bureaucrats. The major reason for the failures in this field is that the systems introduced by the people in power were either imported or were designed to perpetuate their personal rule. The nation's needs and aspirations were completely disregarded. The Army cannot be used as a universal panacea for the politicians' misrule. It has its own duties to perform. Unfortunately, the failure to find a viable political system has resulted in a climate of constant political instability that affects the growth of national power. A workable political system will have to be introduced which would be generally acceptable. If it is to have a chance of success the new system must be structured in conformity with the culture, traditions, needs, ideology, and aspirations of the people. It is heartening that the regime in power is seriously working to overcome this impasse and to meet these essential requirements. A viable political system, when introduced, will not only remove the serious vulnerabilities of this transitory system but will bring about a sense of permanency and greater stability inside the country.

National Cohesion

Although the lack of national cohesion in Pakistan is often referred to as a serious vulnerability in nation building, it certainly is not as serious a liability as outsiders usually imagine. Each of the four provinces has its own language but everyone is united by one common language, culture, religion, and national aspiration. The most serious threat to national cohesion is the economic disparity among the provinces. This complex problem will require large resources to eradicate. The government has diverted some funds and efforts in this direction but at present the resources available are insufficient for the funding necessary. The danger exists that outside forces may exploit this situation for their own advantage if the problem is not addressed immediately.

Another serious problem is created by the large numbers of Afghan refugees flooding into Pakistan. Today, their number is estimated to exceed 3 million. They are presently being looked after by Pakistan and the world community. Besides creating a security problem for Pakistan their presence is certainly affecting the economic, social, and political atmosphere of the country. The Afghans' prolonged stay will provide them with an excuse to settle down on a permanent basis and to evolve as another entity or a social group. Although hospitality is an essential element in Pakistan's culture, this type of situation may not be acceptable to the native people indefinitely. The Afghans' extended sojourn will strengthen them enough to be partners in any future negotiation and may give them the potential to exercise a greater influence in the policies of the government of Pakistan.

The Economic Situation

Pakistan has been struggling to achieve a growth rate commensurate with the nation's requirements. However, performance in the past—except for a few boom periods—has not been very satisfactory. Besides suffering from a dearth of resources the country has suffered from natural calamities, poor economic planning, bureaucratic mismanagement, a fast-growing population, and three wars in a short span of history. The present government has been working meticulously to roll back many of the ills of the past yet there remains a lot of ground to be covered. However, the performance of the last few years has shown steady progress and has helped create some confidence in this field.

The recent census shows good progress in the rate of industrial and agricultural expansion. The public experienced fewer shortages in food staples and Pakistan's poverty and unemployment figures compare favorably with most of the sub-continent. Per capita GNP is also considered to be the highest in the sub-continent. However, there are some factors which continue to have an adverse effect and need to be mentioned. First, is the problem of energy imports which alone consume about two-thirds of Pakistan's merchandise export earnings. Second, is the population growth rate which certainly neutralizes the positive effects of economic growth. Third, is the burden of the influx of over 3 million refugees. Although the world community has been helping to finance their support, Pakistan's share from its meager resources is not unimpressive. Fourth, is the government's intention to contribute large amounts in the development field to remove disparities amongst the less privileged provinces and areas. This is a political requirement which has to be met immediately, even if it takes away funds from other ongoing projects. Fifth, is the future of foreign remittances—which, in the past, have made a considerable contribution to the economy—in view of dwindling demands for manpower in foreign countries. Last, but not least, is the expenditure on the maintenance of the armed forces in order to meet the security needs of the country. All these factors go to make a complex economic situation for a country which has very few resources and far too many commitments.

As one can see from the brief review of Pakistan's internal situation the problems faced by Pakistan at this point are many and complex, some resulting from domestic difficulties and others sparked by outside powers. However, in order to survive despite the existing complications it is imperative for Pakistan not only to promptly overcome these difficulties but also to present a stable and strong front both to endure as a nation and to provide stability to the freedom-loving world community.

VI

CHALLENGE AND RESPONSE

The Soviets are in Afghanistan as a part of a grand design to fulfill their historical dreams; by coincidence the design also serves their current global interest. Afghanistan by itself is no attraction of any consequence to the Soviets, but the invasion of Afghanistan has provided multiple strategic options in the direction of the Soviets' ultimate objective of a grand power play. In order to maintain the stance of a superpower and to keep up its image as the leader of the communist world the USSR feels compelled to achieve success in Afghanistan as speedily as possible and then to probe forward.

If it were successful in its southward maneuver, the USSR would make major gains. It would reduce its lengthy and tenuous line of communication from the Atlantic to the Pacific and make the line more secure against hostile action. The Soviets would be able to reduce the reaction time to Atlantic and Pacific bases with consequent repercussions in the Mediterranean.

Accomplishment of the Soviet design would enable the USSR to impact dramatically on energy supplies to Western Europe and Japan; outmaneuver China; restrict US activities in the Indian Ocean and impose caution on US deployments elsewhere; completely eliminate any freedom of action for Pakistan, India, and countries around the Persian Gulf; and wield greater influence in Asia, the Middle East, and Africa to gain unprecedented economic and political advantages.

The influx of three million refugees from Afghanistan is beyond the capacity of a country like Pakistan to host for long. Besides causing a drain on existing resources, the Afghans' presence has other implications, which are gradually surfacing, in the field of economic, political, social, and security matters. The problem of the Afghan refugees, therefore, is an important factor to be understood. It calls for a speedy solution before Pakistan becomes, as some have already called it, Afghan-Pakistan.

It would be difficult to say when, how, and in what environment the Soviets would hope to accomplish the final goal. Nevertheless, the Soviets are oriented toward the chosen direction and unless their stakes in Afghanistan rise immensely and/or they are strategically outmaneuvered elsewhere they will sooner or later attempt the final thrust.

With the elimination of Afghanistan as a sovereign country, Pakistan has assumed the unenviable status of a front-line state with all the implications that carries. It has become the neighbor of a superpower. Moreover, the clash of the superpowers' interests in the region further complicates the situation, and places Pakistan in the crossfire of their rivalries and ambitions.

Geography has placed Pakistan in the path of the Soviets. As a result, it is under serious threat of intimidation. The situation is further complicated when Pakistan's relation with another major neighbor, India, is judged in its historical perspective, and Pakistan's internal problems are given due weight. A stable and strong Pakistan is an absolute necessity to check the Soviets' ambitions. Essential ingredients for Pakistan's stability lie in its economy, political system, national cohesion, and very well-equipped armed forces. Another important factor is the future of Indo-Pakistan relations. India must be made to understand the changed situation and accept that a strong Pakistan will only help India and the region to face the common threats in the area. Regional stability would be improved if Pakistan and India were able to resolve their age-old problems and decide to live in peace and harmony. A stable region would by itself present a bulwark against aggressive Soviet designs.

In conclusion, it is essential to mention that Pakistan, which is caught up in the quagmire resulting from the Soviet invasion, continues to show great resolve in meeting the challenges facing it. In general terms, Pakistan is making great strides in many fields to achieve stability. The new political situation may further help in this regard. In any case, the volatile conditions across the border call for a greater struggle and vigilance and, of course, for the cooperation of all freedom-loving countries of the world.

BIBLIOGRAPHY

Bhargava, G.S., "South Asian Security After Afghanistan," *Defense & Foreign Affairs,* February 1985.

Braun, Dieter, "The Afghanistan Conflict as a Regional Problem," *Asia and Middle Eastern Studies,* Vol. 6, No. 4, Summer 1983.

Harrison, Selig S., "Dateline Afghanistan," *Foreign Policy,* 1980–1981.

Iqbal, Pervaiz, "The Imperative of National Security—A Case Study of Pakistan," *Asia Defense Journal,* November 1983.

Jacobs, C. "Changing Defense Problems of Pakistan," *Islamic Defense Review,* Vol. 7, No. 4, 1982.

Khalid, Zulfigar A., "Pakistan's Security Dilemma," *Asia Defense Journal,* July 1983.

Malik, Hafeez, "The Failure of the French Negotiations on Afghanistan," *Journal of South Asia and Middle Eastern Studies,* Vol. 4, No. 3, Spring 1981.

Malik, Hafeez, "The Afghan Crisis and Its Impact on Pakistan," *Journal of South Asia and Middle Eastern Studies,* Spring 1982.

"Pakistan's Search for a Foreign Policy after the Invasion of Afghanistan," *Pacific Affairs,* Summer, 1984.

"Pakistan Since the Soviet Invasion of Afghanistan," Strategic Studies Institute, US Army War College, 25 January 1982.

"Pakistan's Relation with India and Afghanistan," *Journal of South Asia and Middle Eastern Studies,* Vol. 5, No. 3, Spring 1982.

"Soviet Policy Dilemma in Asia," *Asia Survey,* June 1977.

Tahir-Kheli, Shirin. "Soviet Fortunes of the Southern Tier: Afghanistan, Iran, and Pakistan," *Naval War College Review,* Winter 1979.

"The Arc of Crisis," *Foreign Affairs,* Spring 1979.

The Military Balance 1980–81. (London: International Institute for Strategic Studies, 1980).

The New York Times, 19 September 1979.

"The Pakistan Unrest and the Afghan Problem," *The World Today,* May 1984.

The Security of Pakistan, DTIC, September 1980.

"The Soviet Strategic Emplacement in Asia," *Asia Affairs,* February 1981.

POSSIBLE DEFENSE ROLES FOR THE ARABIAN GULF COOPERATION COUNCIL

Colonel
Abdulaziz bin Khalid Alsudairy
Royal Saudi Air Force

CONTENTS

TABLE

MAP

FIGURE

APPENDIX

I

ENERGY AND RESOURCES

The perception is held widely throughout the world that the strategic importance of the Middle East is inextricably joined to access to the energy resources of the area by the West. The purpose of this paper is to discuss the strategic importance of the region and to suggest some defense arrangements acceptable to the nations of the region. The paper is not intended to be a detailed defense plan nor the outline of a scenario for future hostilities. Instead, it highlights the dynamics of the region and the implications of a collective defense arrangement.

II

THE GULF

No arm of the sea has been or is of greater interest alike to the geologist and archaeologist, the historian and geographer, the merchant, the statesman and the student of strategy than the inland water known as the Persian Gulf.

Sir Arnold T. Wilson,
The Persian Gulf: A Historical Sketch from the Earliest Times to the Beginning of the Nineteenth Century.

The quotation provides a much needed larger perspective for any discussion of the Gulf and its surrounding setting.* The comment, and the book from which it was taken, make clear that the importance

---

* Oxford: Clarendon Press, 1928, p. 1.

of the geographical area predates and is far more basic and enduring than the current energy-related stature of the region, with which much of the West seems to be so narrowly preoccupied, even paranoid at times.

It is also worth noting at the outset that the prevailing tendency in the West to refer to the waterway as the "Persian Gulf" could cause some to predetermine, if not preempt, just who has the more basic claim, role, and association with this strategic waterway. That claim is quite unacceptable to the overwhelming majority of societies in the region as well as to much of the international community.

A far more apt description for this special sea is the "Arabian Gulf," which is what America's friends in the area call it. Still others have sought to get around the diverse sensibilities involved by suggesting the name should be the "Islamic Gulf." That would reflect the overwhelming reality which pervades the entire setting. The concept is more relevant to the stability and security of the region than energy considerations, development, the concerns of the international community, and the vaunted but tenuous regional projections of the superpowers.

The Islamic nature of the entire area is worth reflecting on at the beginning of this paper. For if that commitment is not recognized and respected, there is little possibility of success for any security role that might be devised. This is true for each individual country, whether the scheme involves a present regime or a long-term projected structure within the Gulf, and whether or not any outside power participated. Islamic sovereignty over the area, its people, and institutions defines the core of everything consequential and lasting there—and limits what "role" anyone, or any power, might seek to play in that part of the world.

In a paper presented for consideration in the United States, it may be easiest to refer to the waterway and the surrounding setting simply as "the Gulf". The importance of the area and familiarity worldwide have made such a reference clear and concise for scholars, strategists, and the media in all parts of the globe. Even in the United States, where fifteen years ago mention of "the Gulf" automatically was taken to mean the next-door Gulf of Mexico, it is now widely taken for granted that the phrase refers to the kidney-shaped, roughly

500-mile long body of water between the Strait of Hormuz and the Shatt al-Arab delta of the Tigris and Euphrates Rivers, despite its being over 9,000 miles distant.

Over 60 percent of the world's proven petroleum reserves lie in and around the Gulf. That percentage is fairly certain to increase as other important sources decline. The North Sea and Alaskan pools and proven US reserves, after cresting at the beginning of the 1970s, are likely to continue their slow, steady drop through the 1990s. Similarly, Soviet production is already having to turn to higher and higher cost Siberian fields that are increasingly difficult to drill. A number of other present sources in South America, Southeast Asia, Africa, and elsewhere will be draining dangerously lower over the course of the coming two decades.

Even amid the current global oil glut and the unsteady spot oil pricing situation, it is well to keep in mind the report released in October 1984 by the US Congressional Office of Technological Assessment. This noted that both domestic American and international dependence on overseas oil is likely to grow by the 1990s, despite increased conservation and fuel switching. In terms of the next several years, the need to replace most US oil imports would have severe economic consequences, including a projected 6 percent drop in the gross national product. There would be an approximate doubling of oil prices even while using America's strategic petroleum supplies.

The significance of the Gulf's energy, however, lies not only in its huge volume and longer-lasting availability, but in the fact that some of the Gulf states—particularly Saudi Arabia—are politically and economically able to increase production fairly quickly and substantially so as to meet emergency shortages or increased longer-term international demand. Equally important, they also are able to reduce their production markedly to moderate gluts and accompanying downward oil pricing shocks. All this is crucial to the international economy, keeping the shifting economic balances among key parts on a steady course and providing time for unavoidable adjustments to be made. The Gulf has a major role to play, not only as a source of energy, but in reinforcing the international economic and financial stability essential to real security in the world.

These brief comments on the Gulf's energy significance should not, however, mislead one to conclude that the area's oil importance is of any comparable consideration with the Islamic nature of the area. It bears repeating that this Islamic nature is the defining and overriding dynamic for all purposes, including defense efforts.

The Superpowers

The strategic interests held by the United States and the Soviet Union contribute in turn to the strategic importance of the Gulf to the international community. The statement of US objectives in the region reflects the perceived strategic importance of the area. These objectives are to preserve and protect the independence of states in the region (including both Israel and friendly Arab nations) from aggression and subversion; to help secure a lasting peace for all the peoples of the Mideast; to prevent the spread of Soviet influence and the consequent loss of freedom and independence it entails; to protect Western access to the energy resources of the area, and to maintain the security of key sea lanes to this region.

A major outside power, Britain, was involved in the region's security arrangements from the late eighteenth century until 1971. In the later years, the United States became Britain's key partner and subtle rival. Today, the direct involvement of a major outside power would undoubtedly contribute to instability in the region. This would be especially true if potential arrangements included landbasing or substantial presence of superpower forces.

III

THE GULF COOPERATION COUNCIL

The geography of the Gulf region, the territorial positions, common borders, and population centers would suggest that a collective self-defense arrangement for the conservative Gulf states might be in the best interest of all concerned. The following brief description of initiatives toward creating collective security arrangements serves as an introduction to the Gulf states which will be discussed next.

Upon reaching acceptable border settlements—although most borders are not formally demarcated—Saudi Arabia and Kuwait joined together to assist Bahrain, the United Arab Emirates (UAE), and Qatar in the management of their economies and the development of their resources, petrochemical and aluminum industries.

The creation of the Gulf Cooperation Council (GCC) in February 1981 provided the infrastructure upon which a collective defense arrangement can be based. The GCC currently includes Saudi Arabia, Kuwait, Bahrain, the UAE, Qatar, and Oman and was formed to provide for mutual defense, economic, educational, industrial, agricultural, political, and security cooperation. The Council is comprised of a Supreme Council (the six heads of state of the member nations), and a Ministerial Council, consisting of the six foreign ministers of the member states and the six defense ministers, and the chiefs of staff of the military establishments.

The GCC states have taken giant steps toward modernizing and improving their military forces and cooperation (see table, p. 157). The agreements and understandings that have been achieved by the GCC to date have been impressive both in nature and scope. These agreements and understandings are documented in civil aviation, standardization of educational programs, exchange of information at all levels, standardization of customs procedures and tariffs, establishment of joint economic venture arrangements, harmonization of development programs, and countless other cooperative efforts. While these efforts are clearly diplomatic and economic in nature, security arrangements are of paramount concern. To this end, efforts in progress or in the planning stage include a regional Rapid Deployment Force, standardization of close air support procedures, standardization of command control communications and intelligence, coordination of the regional air defense network, and recruitment and training of military personnel.

GCC Members

Bahrain Bahrain is made up of a group of islands in the Gulf, midway between the tip of the Qatar Peninsula and the mainland of Saudi Arabia. The main island of Bahrain has an interior plateau 30 to 60 meters in elevation with a hill, Jabal Dukhan, rising to 135 meters above sea level, the highest point on any of the islands.

Approximately 66 percent of the indigenous population came from the Arabian Peninsula and Iran. Islam is the major religion with the Sunni sect predominating in the urban centers and the Shi'a sect in the rural areas. The estimated population is 400,000.

Road networks are limited and no rail system exists. Cross country transportation is fair, with no major obstacles such as forest, rivers, or large urban areas. The total area of Bahrain is 676 square kilometers (260 square miles). Bahrain is served by an international airport at Manama, the capital.

The Bahrain Defense Force numbers 2,800 personnel and consists of an Army, Navy, and Air Force. The organization and military equipment belonging to Bahrain and the other Gulf states are summarized in the Appendix, together with details of other armed forces of the region.

Kuwait Kuwait is located in the northeastern corner of the Arabian Peninsula, bounded on the north and west by Iraq, on the south by Saudi Arabia, and on the east by the Gulf.

The population of Kuwait is approximately 1,750,000 people who are primarily Arab in origin. Most native Kuwaitis are Sunni Muslims and about 30 percent is Shi'a.

The road network is limited to one principal north-south road and one east-west road. Kuwait City, the capital, is the country's major seaport.

The defense forces of Kuwait total approximately 12,500 personnel. The Army, their largest service, is 10,000 strong; the Navy is primarily a coast guard; and the Air Force has an intercept and ground attack role.

Oman Oman is located in the eastern part of the Arabian Peninsula. Its land borders with Saudi Arabia and the U.A.E. are still undefined, while the border with South Yemen is in dispute. The eastern side borders on the Gulf of Oman and the Indian Ocean. Oman's territory includes the tip of the strategically important Musandam Peninsula which overlooks the Strait of Hormuz, the passageway for much of the region's oil production.

The population is estimated to be approximately 1 to 1.5 million people. About one-third of the population lives in Muscat, the capital, and the Batinah coastal plain to the northwest of the city.

The road network is limited with one principal road which runs from Muscat southwest to the border with South Yemen. An international airport is located 36 km west of Muscat. Floods make transportation difficult during monsoons.

Omani defense forces number over 20,000 personnel. Military service is voluntary. The army is organized with two brigade headquarters to which battalion size units are assigned when tasked. The Navy performs coastal protection and limited amphibious operations from bases at Muscat and Ghanam. Air force capabilities include ground attack, reconnaissance, transport, and counterinsurgency operations.

Qatar Qatar occupies the peninsula which juts out into the Gulf. It borders the U.A.E. and Saudi Arabia in the south. Bahrain lies to the northeast.

The population of approximately 270,000 residents comprises almost 80 percent expatriates from Iran, India, and Pakistan. The Qataris are mainly Sunni Muslims of the Wahabi sect.

The terrain is mostly flat and barren, covered with loose sand and gravel, and relieved only in the western areas by low ridges. A lattice-like road network connects most major towns and the capital, al-Dawhah (or Doha), which has an international airport.

Qatar maintains a military establishment of about 6,000 volunteers.

Saudi Arabia Saudi Arabia is geographically the largest of the GCC states occupying about four-fifths of the Arabian Peninsula. Bordered on the east by the Gulf and the Red Sea on the West, Saudi Arabia has common borders with Qatar, the U.A.E., Oman, North Yemen, and South Yemen to the east and south, and on the north with Jordan and Iraq. The topography is mainly desert with the terrain sloping gently to the east from the mountains near the Red Sea.

The population is estimated at about 12 million people. Observers frequently connect the country's stability to its Islamic heritage. About 95 percent of the population has settled in urban centers.

The road network is limited and domestic travel between major cities is supported by daily scheduled flights. Airports serve nearly all major urban areas.

The Saudi military service is the most advanced of the GCC member states with over 50,000 personnel, and the Royal Saudi Air Force is one of the most advanced air forces in the world. A two fleet navy is required by the two contiguous waters, the Red Sea and the Gulf. A National Guard of about 25,000 can field forty infantry battalions.

United Arab Emirates The U.A.E. is bounded on the north by the Gulf and shares common borders with Qatar, Saudi Arabia, and Oman. The climate is hot and dry with little rainfall.

Fewer than 20 percent of the population of 1.3 million are U.A.E. citizens, the rest of the population being made up of Palestinians, Egyptians, Yemens, Omanis, Iranians, Pakistanis, Indians, and Europeans. Most of the indigenous people are Sunni Muslims.

The terrain is primarily flat, barren coastal plain which changes to rolling sand dunes inland, eventually merging into an uninhabited wasteland. Paved roads link the seven emirates which make up the U.A.E.

The Union Defense Force and the armed forces of the U.A.E. merged in 1976, forming a force of over 40,000.

The Total Force Aggregate

Assuming some measure of comparability, the total force aggregate of the GCC becomes a more formidable deterrent than would be any single armed establishment in the region. This amalgamation of national forces into a supranational force must inevitably meet any regional threat to the GCC membership. In all likelihood the existence of the supranational force serves as a deterrent in a number of scenarios. The unfortunate reality is that these forces even when grouped together, are woefully inadequate.

IV

THE GULF SETTING

Ideally, the GCC would include among its membership all the nations of the Gulf region. Today, this is far from being the case and the present alignment of nations adds nothing to the stability of the

GCC COMBINED FORCES

Country:	OMAN	KUWAIT	SAUDI ARABIA	QATAR	UAE	BAHRAIN
Artillery bdes.	1	1	2	(1Bn)	1+(5Bn)	(AFV sqn)
Infantry bdes.	2+	2(Mech.)	3(M)+1	1+(5Bn)	3(9Bn)	(1 Bn)
Artillery regts.	3	–	2+(5Bn)	(1 bty.)	1	(1 bty.)
Guards/s.f. bde.	1/1recce	1/1	1	1		
Tanks:	18	240	450	24	196	—
AFV:	36	180	370	40	90	36
APC:	—	325	1000+	161	330	—
AIR FORCES:						
Ftr. Sqn.	1–2	1	3–4	(1)	2	—
Ftr/att.Sqn.	1–2+1	2	3	(1)	(1)	—
COIN:	1–2	1	—	—	1	—
Hel.AT:	—	23	—	—	13	—
Transport:	24	9	36	9	22	12
Tpt.ac:	24	5	70	3	24	—
NAVY:						
FMB:	3	—	13	2	6	(2)
FAC(g):	4	(2)	4	6	6	2
Patrol:	4	15	—	38	6	—
LC:	6	6	12	—	—	—
Air defence	28	1/Hawk 1 Bn	1/Hawk 16+2 bty.	Tigercat	1/Hawk Rapier,	Rapier Crotale

Source: Defence Update 44

area. The GCC member states are far from being isolated from other potentially hostile states. An examination of the map of the Gulf region reveals that in addition to the GCC members, Iran and Iraq are contiguous to the Gulf, and Israel, Ethiopia, Syria, North Yemen, and South Yemen all lie within close proximity to both the Gulf and the GCC states. With the exception of Ethiopia, all these potential adversaries have overland routes into one or more of the GCC states. A closer look at each of the states in the region, examining its geography and military posture, is necessary to set the stage for further discussions of defense matters. The physical geography of the region is relatively homogeneous throughout the Gulf region. As regards climate, it is generally extremely hot in summer and, because of the proximity to the Gulf, tends to be humid.

Threats to the Gulf

To understand the countervailing forces in the Gulf region it is necessary to understand the potential threat to the GCC countries, collectively and individually. The range of conflict which could result from one hostile action is staggering. The possible involvement of the superpowers takes on an added dimension in view of the stated objectives to maintain access to the Gulf's energy resources. Within the region, conditions exist which could draw GCC members into untenable positions in an effort to maintain a semblance of national integrity. An assessment of the threats that the GCC states face requires dividing the threats into theoretical categories—internal, inter-regional, and extra-regional.

Internal Threats

Few threats are exclusively internal to the GCC. Even situations which are internal in origin inevitably become international in nature as regional or external powers seek to exploit these situations for their own purposes. Internal threats include domestic unrest, revolution, coups, secessionist movements, and civil wars. This exploitation and promotion of domestic unrest and revolution by outside interests make it virtually impossible to separate internal threats from external ones.

Recent examples of internal threats have occurred within the GCC or within the region. The Islamic revolution in Iran, though not directly involving a GCC state, was nonetheless a great shock to the Gulf region. The overthrow of the seemingly secure and powerful Pahlavi regime by a popular revolutionary movement had obvious implications for other regimes in the region with similar cultural, religious, ideological, and governmental makeup. Most of them face similar problems, stemming from rapid socio-economic and cultural changes fueled by vast oil wealth. The large-scale foreign immigration which has placed the native population of three of the GCC states in the minority could be destabilizing. Such conditions have historically led to situations of unrest with the foreign population alleging unfair treatment.

The attempted coup in Bahrain in December 1981 was clearly the catalytic event in the history of the GCC which led the member states to agree on the need for defense and security cooperation. The Bahrain coup attempt and the Dhofar insurgency in Oman from 1965 to 1977 are examples of threats which were originally internal but were later discovered to have been exploited by external powers. The Bahrain coup attempt had Iranian involvement. The "Islamic Liberation Front" acted as the instigator, recruiter, and paymaster of the plot. The Dhofar insurgency, originally tribally based, was taken over at its height by Marxist leaders backed by the PDRY, acting as Soviet surrogates.

Threats like these can be dealt with by the defense and security establishments of the GCC members most efficiently with some degree of cooperation among the member states.

Regional Threats

While it is impossible to predict the intentions of regional powers beyond the GCC boundaries, it is important to establish a viable evaluation of the capabilities of those forces in close proximity to the Gulf. This is not to imply that these powers intend to take any action involving the use of their military potential. In any conflict involving GCC members movement into the area of the Gulf itself is a prerequisite. Therefore, the broad threat axes will be reviewed and then the regional states that might use them.

The Threat From the North

Kuwait and the Strait of Hormuz would be directly threatened by an approach from the north. Since this approach is completely overland, large ground forces could be employed in any effort to assault the Gulf. The mountainous terrain of Iran and eastern Iraq, while difficult to negotiate, would make the task of moving opposing forces into that area nearly impossible, especially in the case of large ground formations. The western edge of this axis, along the Iran-Iraq border, leads to Kuwait and could initially compartmentalize and limit the physical threat if used by a rapidly moving and aggressive adversary. Concentration of GCC forces would require relocations of the highest magnitude, from the most distant force location. This axis can support the most formidable of ground forces.

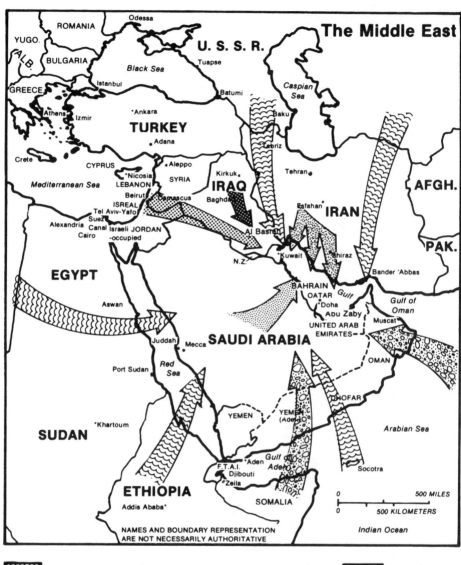

The Middle East

Axes of Threat to the Gulf

Soviet	Soviet
U.S. (Plus Britain, France)	U.S. (Plus Britain, France)
Israeli	Israeli
Iranian	Iranian
Iraqi	Iraqi
Saudi	Saudi

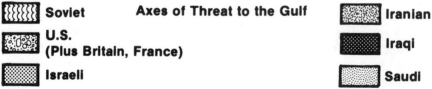

The Threat From the Northwest

The axis from the northwest approaches along the Saudi-Iraq border trace. This approach could isolate Kuwait from the other GCC members if a sustained drive by an attacker were to continue to the Gulf. This avenue confronts the largest force of the GCC group, the Saudi armed forces.

The Threat From the West

This approach is complicated by the presence of the Red Sea which would dictate an amphibious operation to gain access to the Arabian Peninsula. To arrive at the Gulf would require a ground force to transit the entire width of the Arabian Peninsula. Attacking forces would have their backs to the Red Sea and would not have the benefit of a relatively safe staging area across a contiguous international border.

The Threat From the South

On this axis one or both of the Yemens could be used as a beachhead. This eases the requirement for an assault landing which would be required for landing directly in one of the GCC states. The terrain that has to be crossed to reach the Gulf is very harsh.

The accompanying map summarizes these axes and suggests possible users of the approaches. The lines from the east, south, and west would surely require support from a sizeable amphibious force. The superpowers routinely maintain naval forces in the Mid-East, as the map also indicates.

V

THE OTHER MID-EAST STATES AND THE UNITED STATES

The reality of the current geopolitical situation dictates that some states are refused membership in the GCC or choose other arrangements for national development and security. The Figure "The Present Geopolitical Context" of the Gulf lists some of the current

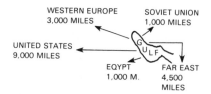

THE PRESENT GEOPOLITICAL CONTEXT OF THE GULF

GULF SOCIETIES ARE LINKED TO THE STABILITY OF THE MIDDLE EAST AS A WHOLE AND ESPECIALLY TO ARAB-ISRAELI CONFLICT BY BLOOD, RELIGION, CULTURE, ECONOMICS, HISTORY, MODERN COMMUNICATIONS

WHAT STRENGTHENS, OR WEAKENS MIDDLE EAST STABILITY STRENGTHENS, OR WEAKENS, THE GULF'S BASIC STABILITY AND SECURITY.

THE RED SEA FRONTIER CONTAINS MAJOR SEEDS OF INSTABILITY COMPLICATING THE GULF STATES' SITUATION — AS OUTSIDE ATTEMPTS TO SUBVERT SUDAN ARE ADDED TO INTERNAL DIVISIONS AND CONDITIONS THERE.

THE ETHIOPIAN RELATION-SHIP WITH SOVIET AND CUBAN TROOPS THERE PROVIDE A MAJOR BASE FOR REGIONAL LEVERAGE.

MIDDLE EAST & APPROACHES

THE STRAIT OF BAB-EL MANDEB IS ONLY 20 MILES WIDE; THE ECONOMY OF THE GULF AS WELL AS EGYPT AND A NUMBER OF OTHER COUNTRIES ARE DEEPLY AFFECTED BY THIS STRAIT BEING MANIPULATED AS A STRATEGIC CHOKEPOINT

"SUPERPOWER" STRONG POINTS ARE ALL ALONG THE SOUTHERN FRONTIER OF THE ARABIAN PENINSULA — THE SOVIETS AT SOCOTRA AND ADEN, AND THE US IN MASIRA AND OMAN; BOTH IN THE ARABIAN SEA AND THE INDIAN OCEAN.

THE COLONIAL BURDEN WAS LIFTED FROM THE UPPER GULF ONLY IN LAST HALF CENTURY AND FROM THE LOWER GULF IN THE LAST DECADE AND A HALF...FROM SOME OF THE HORN OF AFRICA IN THE SAME TIME FRAME. POLITICAL INSTITUTIONS, ECONOMICS, AND SELF IDENTITY ARE STILL DEVELOPING AND THESE COUNTRIES ARE SELF-CONSCIOUS OF THEIR STATUS.

THE SOVIET UNION IS ONE HOUR'S FLYING TIME FROM THE GULF — AND THE SOVIETS' MAJOR AIRLIFT CAPACITY OVER THE GULF AS FAR AS ETHIOPIA HAS BEEN PROVEN.

THE SOVIET INVASION OF AFGHANISTAN IS NOW IN ITS 5TH YEAR.

PAKISTAN-INDIAN TENSIONS SEETHE JUST OVER HORIZON.

THE IRAQ-IRANIAN WAR IS NOW IN ITS 5TH YEAR.

THE IRANIAN REVOLUTION'S VERSION OF ISLAM IS BEING EXPORTED BY RADIO TO ALL THE OTHER GULF STATES.

ALL GULF SOCIETIES HAVE SHIITE POPULATIONS — AND SOME HAVE IRANIAN AS WELL AS OTHER RESTIVE MINORITIES.

OVER 60% OF THE WORLD'S OIL RESERVES ARE AROUND THE GULF.

THE STRAIT OF HORMUZ IS 35 TO 60 MILES WIDE AND ONE OF THE GLOBE'S KEY CHOKEPOINTS.

THE GULF IS SUFFICIENTLY SMALL THAT MANEUVERING IN IT BY LARGE NAVAL SHIPS IS NOT EASY

VARIOUS STATES IN THE AREA ARE ALL GENERALLY STILL SEEKING TO DEVELOP AND SECURE THEIR IDENTITIES AGAINST OUTSIDERS AND KEY NATIONS WITHIN THE REGION.

FREE WORLD ECONOMIC AND SECURITY COHESIVENESS DEPENDS ON THE GULF BEING OPEN AND FUNCTIONING!

realities which describe the geopolitical context of the Gulf. With this in mind, the states of the broader Gulf setting which do not embrace the GCC are described from the perspective of military potential. This is not to say that these nations are considered adversaries or enemies, only that because of their close proximity or because of conditions beyond their immediate control a greater awareness on the part of the GCC is called for. Ethiopia, Iran, Iraq, Israel, North Yemen, and South Yemen will be discussed. Only the military posture will be included since it is likely that any Gulf conflict would take place in the territories already described.

Ethiopia

Ethiopia is currently engaged in conflict with a number of guerrilla groups including Eritrean, Tigray, Somali, and People's Liberation Fronts. Ethiopia's relationships with the Soviets and Cuba suggest that outside influence could be a factor in any involvement in the region.

Iran and Iraq

The Iran-Iraq conflict continues, adding to the instability of the Gulf region. A threat exists to all who use the Gulf's oil and these two nations have the military power, geographical location, population, and revolutionary fervor to pose a serious threat to the GCC. This threat is currently obscured by the prolonged stalemate of the war in which Iran and Iraq are engaged. Victory by either side in the Iran/Iraq conflict would bring disastrous results for the GCC. Iran would probably demand compensation if not subjugation from the GCC states. Victory by Iraq could cause Iran to carry out, in desperation, the threat to close oil exports through the strategic Strait of Hormuz.

Israel

The Israeli presence in the occupied lands and the unresolved Palestinian problem are the most destabilizing factors in the region. Israeli forces are in the process of pulling out of Southern Lebanon, which they have occupied since 1982. Israel maintains a special relationship with the US; however, its defense policy is very changeable. Hence, the measure of the threat is related to the extent that Israel ignores US warnings on those occasions when it feels its security needs justify military actions. Israel thus creates, from time to time, very hazardous and threatening situations for the whole area.

North Yemen (Yemen Arab Republic)

North Yemen is listed in view of its proximity to the Gulf and its border with Saudi Arabia.

South Yemen (Yemen: People's Democratic Republic)

South Yemen has common borders with Oman and Saudi Arabia. South Yemen could provide entry to the Arabian Peninsula for forces landing from the east. North Yemen's stability is threatened by the attacks by South Yemen. While, to date, attacks have not been clearly sponsored by the Soviets, large scale deliveries of Soviet arms indicate a distinct possibility that this was covertly the case.

Analysis

The aggregate force of the GCC member states falls short of numerical superiority over some of the non-member states' forces. Assuming that it is unlikely that all these states would band together for a simultaneous assault on the Gulf, the collective defense arrangement afforded by the GCC effort would be a workable solution for a temporary delay and defense.

The geographic locations of the states just described dictate the approach avenue most likely to be taken in order to arrive at the Gulf. Ethiopia's probable routes would be from the west and south. Considerable amphibious support would be required. Both Iran and Iraq would be likely to favor the routes from the north while Israel would likely take the northwest route. The Yemens would obviously use the southern approaches.

The Threat From the United States

The threat posed by the United States is a many-faceted problem that must be considered from the aspect of the probable results for the Gulf region. While the US objectives in the region may be well-intended, the problems that could be created lead to a severe restriction, on the part of the apprehensive Gulf states, of US involvement in the military defense of the region. This is not to say that there is no role for the US in the region—but the likely reaction to the presence of US forces must be considered.

First, introduction of US forces into the region would most certainly cause retaliation by the current regime in Iran and the introduction, on a larger scale, of Soviet forces at some other point in the region. The Iranian effort could be expected to take the form of terrorist activities against US forces—activities which inevitably would spill over to the local population. Pro-Khomeini elements could also be expected to attack GCC members in retaliation for the GCC allowing the stationing of US forces in the region.

Additionally, introduction of US forces could cause a split between internal factions who support the necessity of US forces, and those who denounce a US presence, either because of their leanings towards the Soviets, or from a desire to maintain autonomy. Finally, the most likely result would be a Soviet countermeasure. This could be in the form of direct Soviet intervention or surrogate action through some other country eager to enter the region but without the wherewithal to do so without Soviet backing.

VI

ALTERNATIVES FOR THE DEFENSE OF THE GULF

Now that the participants have been identified and the scene described, possible approaches to the defense of the Gulf need to be considered. Historically and analytically, there appear to be three basic models which could be used for structuring security arrangements for the Gulf area and selecting the principal roles for its defense.

(1) *An Overlordship of the Area:* This first model describes dominance by a major outside power such as Britain exercised from the late eighteenth century until 1971. Far earlier and for a much longer period, the overlordship of the Gulf was exercised by the Islamic civilization which began in western Saudi Arabia and was ultimately centered in Baghdad. Somewhat similarly, the Persian

Empire, roughly 2,500 years ago, sought overlordship of this inland sea, although it never took control of the western shores of the Gulf.

(2) *An Externally Structured Arrangement:* The second basic model for structuring the security arrangement for the Gulf (and for defining the roles of the various societies there) and relevant outsiders, is most readily suggested by the "Twin Pillars" concept of Dr. Henry Kissinger and other US theorists in the 1970s. They sought to rely on Iran and Saudi Arabia as the key positions in the immediate area. This approach was certainly never adopted by Saudi Arabia or other societies in the region. In fact, it was more a descriptive rationalization in Washington than an actual operational model. It does, however, illustrate an alternative approach to Gulf security.

(3) *An Internally Structured Defense of the Area:* This third model for structuring the defense of the area is a broad-based concept for the countries there. In practical terms, political, economic, and defense coordination is institutionalized by as many states as have closely compatible interests which can be furthered by collective efforts, but with no single state visibly dominating the group.

In addition to the three approaches above, there are undoubtedly other possibilities, or at least more elaborate refinements of the three formulations summarized here. For comparative purposes, these three provide a manageable frame of reference for reflecting on the overall problem.

History has shown that an overlordship has never been an acceptable arrangement for the Gulf region. Such an arrangement is doomed to failure whether the overlord power is conspicuous by its presence or its absence. The reality of cultural differences easily strains relationships and the indigenous population soon begins to feel oppressed.

The "Twin or Multiple Pillars" strategy is likewise unworkable if the foundation states are not in agreement with the outside power from the beginning. Selection of "Pillars" smacks of favoritism and would tend to alienate other regional states that do not enjoy the same favors.

The third model, which describes a cooperative arrangement by the regional states, appears more palatable to the regional inhabitants. It is, however, less desirable for outsiders who have a perceived interest in the region. Some middle ground can probably be found where the best interests of all parties are equitably addressed.

Limits on the Acceptability of a US Role Within the Gulf States

Regardless of the model used to structure the security arrangements of the Gulf region, there are limitations to a US role that would be acceptable in the Gulf setting. Any landbasing or substantial presence in the area by a superpower would disequilibriate and almost certainly contribute to instability in the long run. The introduction of Western military forces would almost inevitably cause very sharp strains at the cultural and human level. Such a presence would result in charges of neocolonialism. Given the current balance in international relations, if one superpower were to have an explicit major role in the Gulf, the other would certainly manipulate its way in at another point somewhere in the region.

In the last 12 years, America has hardly proved to be a credible partner in security or in meeting the major problems of the developing world, especially the Islamic and Arab world. This is proven by the US withdrawal from Southeast Asia, US unwillingness to aid the faltering Shah of Iran and, more recently, in the pullout from Lebanon. US unreliability also is seen in US Congressional opposition to the Reagan administration's sale of defense arms and missiles to Kuwait, Jordan, and other Arab states. It is seen, likewise, in the unsuccessful Congressional resistance to the Carter and Reagan administrations' sales of the F–15 and AWACs to Saudi Arabia. Finally, the 1984 Democratic Party platform provision that is opposed to ever committing US forces in the Gulf, even in the event of an extreme crisis there, appeared particularly ominous to the states of the region.

VII

A REGIONAL DEFENSE

A comparison of the GCC members, either individually or on the aggregate level, shows the inherent weakness from a military perspective. The relatively small military establishments dictate some form of cooperative effort. As compared to many of their potential adversaries in the Mid East, the GCC member states are less populous and, accordingly, possess smaller defense organizations. To arrive at a satisfactory cooperative defense arrangement, it is necessary to realize full military and strategic integration which would, of course, require adequate time, concentrated study, and continuous consultation. Because of the special circumstances of the region, and of the GCC states in particular, joint defense would necessitate the introduction of advanced and effective weapon systems in addition to the adoption of joint policies on arms purchases, military training, and a data system which would ensure the optimum use of available capabilities of the GCC member states.

Key defense areas in which the GCC is placing significant emphasis include early warning systems; command, control, and communications (C3) systems; and a rapid deployment force. Supporting activities to accomplish full integration of these areas include joint purchasing, maneuvers, and training. Attacks on neutral shipping in the Gulf and intrusions into GCC airspace by hostile combat aircraft point out the need for an advanced early warning system. The C3 system is a vital aspect of any effort to integrate joint military operations and must be capable of supporting forces using advanced equipment. The Rapid Deployment Force would contribute to the stability of the region and show the determination of the GCC members to preserve their identity and sovereignty and their full commitment to defend their interests and protect their natural wealth against all intrusion.

Joint Maneuvers

A number of benefits are being derived from joint maneuvers such as the PENINSULA SHIELD series which started in October 1983. First, joint maneuvers allow the individual states to familiarize themselves with a range of military equipment and to create an awareness of the problems of joint training and equipment servicing. Another benefit is gained from the opportunity to exercise the Rapid Deployment Force. Valuable training is gained in transportation of the force, command and control of ground operations, and coordination of ground support aircraft. Each time the Rapid

Deployment Force exercises, it demonstrates the resolve of the GCC nations to provide for their own defense.

Early Warning Systems

The integration of early warning systems with other defensive systems can be accomplished by using an advanced C3 system. A communications network linking airborne warning systems to ground radars, ground missile batteries, and air and sea forces of the GCC member states is feasible and provides an increase in the warning time of an impending attack. The operation and maintenance of such a system, with its high degree of sophistication, requires a significant amount of training. Its use in an optimum manner neutralizes GCC weaknesses by providing a qualitative edge to overcome the problems that result from the low density of equipment. The PEACE SHIELD C3 system will provide a very real possibility for a coordinated air defense system.

Military Training

The training required to operate and maintain advanced equipment is a matter of vital importance to the GCC, but just as important is the training of the corps of officers who are charged with the leadership of the military establishment. The GCC coordinates military training in the member states by centralizing the military college system. Use of such a centralized system ensures that the GCC member states with small military forces are able to participate in an educational system which has regional sensitivities, without the penalty of a costly operating overhead for a very small number of students.

Training is not limited to the academic setting, but embraces daily operations. It is expected that for a period of time instructors—for the most part technicians familiar with the advanced equipment—will have to come from outside sources. Their role will be to ensure that adequate training is available to maintain the qualitative edge the equipment is expected to provide.

Help From the West

The primary role of the United States in the Gulf is to deter direct Soviet aggression. The dilemma faced by the GCC is that the US presence needed to deter the Soviets also serves as an attraction to the

Soviets in the region. Clearly, if the United States is freed from the unwanted and unnecessary burden of a wide range of defensive regional requirements it could then mount a better deterrence, the task that it alone can do. It is very clear that the GCC opposes the presence of US military forces in the region for a variety of reasons.

United States' assistance is needed most of all in the effort to improve the capabilities of the GCC states in order that they can provide for their own defense, short of direct Soviet aggression. The GCC needs include assistance in the purchase of high quality weapon systems to help to redress the purely numerical inferiority in the region. Assistance in training is also needed to help overcome numerical inferiority to potential adversaries.

While the wealth of the region allows for the purchase of weapon systems, the means of production of these systems is located in the West. Additionally, it is necessary to obtain from the West the training of operators and maintenance personnel as well as assistance in developing the training base needed to sustain the necessary force readiness.

VIII

GEOPOLITICAL REALITIES AND A POSSIBLE SOLUTION

The Gulf is a unique region of the world and its strategic importance is perceived in many ways. To the West, its value derives from its energy resources. However, for the inhabitants of the region, its Islamic nature is the defining and overriding dynamic for all purposes, including defense efforts. No matter what the viewpoint, it is undeniable that the region has intrinsic value worthy of development and deserving of greater security.

The realities of the current geopolitical situation would suggest that the Gulf region is threatened by both the well-intentioned United States and by the ambitions of other regional factions. The nations of the region are woefully limited in military power, primarily because of their relatively small populations. Even with a consolidation of all the forces of the states with common economic and security goals, it is still questionable whether or not they can defend the Gulf without outside help.

Of the models available for the defense of the region, the most undesirable solution would be to station US military forces in the region. This would most certainly attract retaliation from other factions with other perceptions of the value of the region. The most palatable defense arrangement would be one provided by the nations of the region. It is patently clear that the indigenous forces cannot accomplish their goal as individual national forces.

It is the obvious need for external assistance that suggests a role for the United States in a situation where the actual presence of US military forces is not wanted by anyone. The United States could give invaluable support to the efforts of the Gulf states that have entered into a cooperative defense arrangement. Assistance is specifically needed in the purchase of advanced weapon systems which can compensate for the apparent weakness of the Gulf states, due to their limited populations and the correspondingly small military establishments. Training in operating and maintaining the new equipment to ensure the optimum efficiency of the system is another desideratum.

The formation of the GCC attests to the resolve of the Gulf states to establish a viable regional defense. The cooperative efforts undertaken by the military forces of the member states are reflected in the purchase of weapon systems, the initiation of joint maneuvers, joint training, and in the formation of a Rapid Deployment Force.

The overwhelming conclusion of this paper is that the defense of the Gulf must be in the hands of the inhabitants of the region. The United States can best assist regional efforts and at the same time serve its own purpose of continued access to the region's energy resources by supporting the GCC cooperative defense effort.

SELECTED BIBLIOGRAPHY

Amirsadeghi, Hassein, ed., *The Security of the Persian Gulf,* (London: Croom Helm, 1981).

Anthony, John Duke, "The Gulf Cooperation Council" A Paper Presented to the Annual Convention of the Middle East Studies Association of North America, 5 November 1981.

Browning, John, "The Gulf States Regroup," *The Economist,* October 1984.

Cordesman, Anthony, *The Gulf and the Search for Strategic Stability,* (Boulder, Colorado: Westview Press, 1984).

Daniel, Donald C., ed., *International Perceptions of the Superpower Military Balance,* (New York: Praeger, 1978).

George, James L., *Problems of Sea Power as We Approach the 21st Century,* (Washington, D.C.: AEI, 1977).

Halliday, Fred, *Threat from the East: Soviet Policy from Afghanistan and Iran to the Horn of Africa,* (London: Penguin Books, 1982).

Iskandar, Marwan, ed., "Gulf Cooperation Needs More Thought If Inequality is to be Avoided," *An-Nahar Arab Report and Memo,* 7 January 1985.

Iskandar, Marwan, ed., "Gulf States Will Be the Losers If the Iran-Iraq War is Much Prolonged," *An-Nahar Arab Report and Memo.*

Ispahani, Malmaz Zehra, "Alone Together: Regional Security Arrangements in Southern Africa and the Arabian Gulf," *International Security,* Spring 1984.

Kelly, J.B., *Arabia, the Gulf and the West,* (New York: Basic Books Inc., 1980).

Mangold, Peter, *Superpower Intervention in the Middle East,* (New York: Saint Martin's Press, 1978).

Niblook, Tim, ed., *Social and Economic Development in the Arab Gulf,* (London: Croom Helm, 1980).

Noyes, James H., *The Clouded Lens, Persian Security and U.S. Policy*, (Stanford, California: Hoover Institute Press, 1979).

Olayan, Sulliman S. "The Middle East: Not Just Oil," *U.S.-Arab Commerce*, December 1984.

Ramazani, R.V., *The Persian Gulf and the Strait of Hormuz*, (Alphen aan den Rijn, The Netherlands: Sijthoff & Noordhoff, 1979).

Weinberger, Caspar, *Report of the Secretary of Defense to the Congress on the FY 1984 Budget*, 1 February 1983.

Wilson, Sir Arnold T., *The Persian Gulf: An Historical Sketch From the Earliest Times to the Beginning of the 19th Century*, (Oxford: Clarendon Press, 1928).

Wright, Marcus, "Questions of Defense," *Middle East Economic Development*, 16 November 1984.

"GCC: Gulf Defense Cooperation," *Defense & Foreign Affairs Daily*, 26 September 1984.

"GCC: Military College Centralization," *Defense & Foreign Affairs Daily*, 5 November 1984.

"GCC Plans Joint Maneuvers," *Middle East Economic Developments*, 20 July 1984.

"Middle East Military Survey 1984," *Defense Update 44*.

"Negative Aspects of Gulf Forces' Exercise," *Jane's Defense Weekly*, 29 September 1984.

The Arab Gulf Cooperation Council: Threat Analysis and Strategic Implications, *International Defense Intermetrics*, Report No. 670002.

The Military Balance 1984/85, (London: The International Institute for Strategic Studies, 1984).

Special Dossier On The Occasion Of The Fifth Gulf Cooperation Council Summit Conference in Kuwait; KUWAIT NEWS AGENCY (KUNA), November 1984.

APPENDIX
THE ARMED FORCES OF THE REGION

BAHRAIN

	Personnel
Army:	2,300

Organization
1   Brigade:
   1 Infantry Battallion
   1 Armored Car Squadron
   1 Artillery Battery
   2 Mortar Batteries

Equipment
Armored Cars:
   8 Saladin
  20 AML–90
Scout Cars:
   8 Ferret
Armored Personnel Carriers:
 110 M–3
Guns:
   8 105mm light guns
   6 81mm mortars
   6 120mm recoilless rifles
Missiles:
    TOW antitank weapon
   6 RBS–70 SAM

Navy	300

2   Lurssen 45-meter Fast Attack Craft
2   Lurssen 38-meter Fast Attack Craft

Air Force	200

1   AB–212 Helicopter Squadron

KUWAIT

	Personnel
Army	10,000

Organization
1   Amored Brigade
2   Mechanized Infantry Brigades
1   Surface-to-surface Missile Battalion

Equipment
Tanks:
  70 Vickers Mk 1
  10 Centurion
 160 Chieftain MBT
Armored Cars:
 100 Saladin
Scout Cars:
  60 Ferret
Armored Personnel Carriers:
 175 M–113
 100 Saracen
Guns:
  20 155mm SP howitzers
  — 81mm mortars
Missiles:
  — FROG–7
  — HOT
  — TOW
  — Vigilant
  — SA–7 SAM

Navy
6   Lurssen TNC–45 FAC with Exocet
2   Lurssen FPB–57 FAC
47   Coastal Patrol Craft
6   Landing Craft

Air Force	2,000

2   Fighter/Bomber Squadrons
  30 A–4
1   Interceptor Squadron
  17 Mirage F–1C
   2 Mirage F–1B
3   Helicopter Squadrons
  23 SA–342K Gazelle
  12 SA–330 Puma
Transport Aircraft
   2 DC–9
   6 C–100
1   Improved Hawk Battalion

OMAN

Army	*Personnel* 16,500

Organization

2	Brigade Headquarters
1	Royal Guard Brigade
1	Armored Regiment
2	Field Artillery Regiments
1	Reconnaissance Battalion
8	Infantry Regiments (Battalion equivalents)
1	Special Force
1	Signal Regiment
1	Field Engineer Regiment
1	Parachute Regiment

Equipment:

Armored Vehicles:

6	M–60 Tanks
—	Chieftain Tanks
30	Scorpion Recce Vehicles

Artillery:

24	88mm Guns
39	105mm Guns
12	M–1946 Guns
12	130mm Guns
—	60mm and 81mm Mortar

Missiles:

—	Milan Antitank Guided Weapons

Navy	2,000
4	Fast Attack Craft with Exocet:
	3 Province
	1 Brooke Marine
4	Brooke Marine Fast Attack Craft
4	Inshore Patrol Craft
1	Logistic Support Ship
4	Medium Landing Craft
1	Training Ship

Air Force	3,000
2	Fighter Squadrons
	20 Jaguar
	4 T–2
1	Fighter/Recce Squadron
	12 Hunter
	4 T–7
1	COIN/Training Squadron
	12 BAC–167
3	Transport Squadrons
	3 BAC–111
	1 Falcon 20
	7 Defender
	15 Skyvan
	3 C–130H

2	Helicopter Squadrons
	20 AB–205
	3 AB–206
	5 AB–214B
1	Air Defense Squadron
	28 Rapier SAM

QATAR

Army	*Personnel* 5,000

Organization

1	Royal Guard Regiment
1	Tank Battalion
5	Infantry Battalions
1	Artillery Battery
1	Surface-to-air Missile Battery

Equipment

Armored Vehicles:

24	AMX–30 Main Battle Tanks
10	Ferret Scout Cars
30	AMX–10P Mechanized Infantry Combat Vehicles
25	Saracen Armored Personnel Carriers
136	VAB Armored Personnel Carriers

Guns:

8	88mm
6	155mm Self-propelled Howitzers
—	81mm Mortars

Missiles:

—	Rapier SAM

Navy	700
3	La Combattante FAC with Exocet
6	Large Patrol Craft
36	Coastal Patrol Craft
2	Interceptor Fast Assault/Search and Rescue Craft
3	Exocet Coast Defense Systems

Air Force	300
3	Hunter FGA–6
1	T–79
8	Alpha Jet
1	Islander
1	Boeing 727
2	Boeing 707

Helicopters:
   2 SA–342 Gazelles
   2 Whirlwind
   4 Commando
   3 Lynx
Missiles:
   5 Tigercat SAM

SAUDI ARABIA

	Personnel
Army	35,000

Organization
3   Armored Brigades
2   Mechanized Brigades
2   Infantry Brigades
1   Airborne Brigade
1   Royal Guard Regiment
5   Artillery Battalions
18   Air Defense Artillery Batteries
14   Surface-to-air Missile Batteries

Equipment
Armored Fighting Vehicles:
   300 AMX–30 Main Battle Tanks
   150 M–60A1 Main Battle Tanks
   200 AML–60/90 Armored Cars
   350 AMX–10P Mech Inf Com bat
      Vehicles
Artillery:
   100+ 105mm
   50+ 155mm
   — 81mm Mortars
   — 107mm Mortars
Antitank Weapons:
   75mm, 90mm, and 106mm
      Recoilless Rifles
   TOW, Dragon, and Hot Missiles
Surface-to-air Missiles:
   216 Improved Hawk
   48 Shahine
Air Defense Guns:
   M–163 Vulcan 20mm
   AMX–30SA 30mm
   M–42 Duster 40mm

Navy	2,500

2   Fleet Headquarters
1   F–2000 Frigate
4   PCG–1 Corvettes
9   PCG–1 FAC
1   Large Patrol Craft
3   Jaguar FAC
4   Coastal Minesweepers

16 Amphibious Craft
24 Dauphine 2 Helicopters

Air Force	14,000

3   Fighter Squadrons
   65 F–5E
4   Interceptor Squadrons
   15 Lighting
   2 T–55
   62 F–15
4   E–3A AWACS
3   Transport Squadrons
   50 C–130E/H
   8 KC–130H
   2 Jetstar
2   Helicopter Squadrons
   12 AB–206B
   14 AB–205
   10 AB–212
Trainers:
   40 BAC–167

UNITED ARAB EMIRATES

	Personnel
Army	40,000

Organization
3   Regional Commands:
   Western (Abu Dhabi)
   Central (Dubai)
   Northern (Ras al Khaimah)
1   Royal Guard Brigade
1   Armored/Armored Car Brigade
2   Infantry Brigades
1   Artillery Brigade
1   Air Defense Brigade

Equipment
Tanks:
   100 AMX–30
   18 OF–40 Mk 1 Lion
   60 Scorpion Light Tanks
Armored Cars:
   90 AML–90, VCB–40
Mech Infantry Combat Vehicles:
   — AMX–10P
Armored Personnel Carriers:
   30 AMX VCI
   300 Panhard M–3
Guns:
   50+ 105mm
   20 155mm
   — 81mm Mortars
   — 84mm Recoilless Rifles

Missiles:
- — Vigilant Antitank
- — Rapier SAM
- — Crotale SAM
- — RBS–70 SAM

Navy   1,500
6   Lurssen TCN–45 FAC
3   Large Patrol Craft
3   Coastal Patrol Craft
2   Support Tenders

Air Force   1,500
2   Interceptor Squadrons
     25   Mirage 5AD
1   Fighter Squadron
     3   Alpha Jet
1   COIN Squadron
     10   MB–326
Transport:
     2   C–130H
     1   L–100
     1   Boeing 720
     1   G–222
     4   C–212
     5   Islander
     9   DHC–5
     1   Cessna 182
Helicopters:
     7   Alouette III
     8   AB–205
     6   AB–206
     3   AB–212
     9   Puma
     4   Super Puma
     10   Gazelle

ETHIOPIA

                                    Personnel
Army   300,000
Organization
1   Armored Division
23   Infantry Divisions
4   Para/Commando Brigades
30   Artillery Battalions
30   Air Defense Battalions
Equipment
Tanks:
     40   M–47
     150   T–34
     800   T–54/55
     30   T–62
     40   M–41

Armored Vehicles:
     100   BRDM–1/–2 Scout Cars
     40   BMP–1 MICV
     70   M–113 APC
     600   BTR–40/60 APC
Artillery:
     700   75mm, 105mm, 122mm,
           130mm, 155mm
Air Defense:
     —   ZSU–23–57 Guns
     —   SA–2/3/6/7 SAM

Navy   2,500
2   Petya Frigates
7   Osa–II FAC
9   Large Patrol Craft
3   Coastal Patrol Craft
1   Polnocny Landing Ship

Air Force   3,500
10   Fighter Squadrons
     10   MIG–17
     100   MIG–21
     38   MIG–23
     12   Sukhoi
1   Transport Regiment
     14   An–12
     4   An–22
     14   An–26
     1   Il–14
Helicopters:
     32   Mi–8
     24   Mi–24

IRAN

                                    Personnel
Army   250,000
Organization
3   Mechanized Divisions
7   Infantry Divisions
1   Airborne Brigade
2   Special Forces Divisions

Equipment
Tanks:
     100   T–54/55
     50   T–62
     100   T–72
     300   Chieftain
     200   M–47/48
     50   Scorpion
Armored Cars:
     130   Cascavel

Armored Personnel Carriers:
  180 BMP–1
  280 M–113
  600 BTR–40/50/60/152
Artillery:
  1000+ 75mm, 85mm, 130mm
  30   175mm SP
  10   203mm
  65   BM–21 122mm MRL
  3000 120mm Mortar
Air Defense:
  1500 23mm, 37mm, 57mm, 85mm
  —   Hawk
  —   SA–7 SAM
Helicopters:
  —   AH–1 Cobra
  —   CH–47 Chinook
  207 Bell 214A
  35   AB–205A
  15   AB–206

Navy   20,000
  3   Destroyers
  4   Saam Frigates
  2   PF–103 Corvettes
  10 Kaman FAC
  7   Large Patrol Craft
  10 SRN–6 Hovercraft
  3   Minesweepers
  2   Landing Ships
  3   Marine Battalions

Air Force   35,000
  8   Fighter Squadrons
    35   F–4D
    50   F–5E
  1   Recce Squadron
    10   F–14A
    3   RF–4E
  2   Tanker/Transport Squadrons
    12   Boeing 707
    7   Boeing 747
  5   Transport Squadrons
    28   C–130 E/H
    10   F–27
    2   Aero Commander 690
    4   Falcon 20
Helicopters:
    10   HH–34F
    10   AB–206A
    5   AB–212

  39   Bell 214C
  10   CH–47 Chinook
  2   S–61A4
  5 Surface-to-air Missile
          Squadrons
  — Rapier
  25 Tigercat

IRAQ

		Personnel
Army		600,000

Organization
  4   Corps Headquarters
  6   Armored Divisions
  5   Mechanized Divisions
  5   Infantry Divisions
  4   Mountain Divisions
  2   Republic Guard Brigades
  3   Special Forces Brigades
  9   Reserve Brigades
  15   Volunteer Infantry Brigades
Equipment
Tanks:
  4500 T–54/55/62/72
  269   Chieftain T–69
  60   M–77
  100   PT–76
3200 Armored Fighting Vehicles:
      BRDM
      FUG–70
      ERc–90
      MOWAG Roland
      Cascavel
      Jararaca
      BMP
      BTR–50/60/152
      OT–62/64
      VCRTH
      Panhard M–3
      Urutu
Artillery:
      3500 Guns including 75mm,
          85mm, 100mm, 122mm,
          130mm, 152mm, 155mm,
          122mm, 140mm MRL
  19   FROG–7
  9   Scud B
  15   SS–11 120mm, 160mm Mortars
Antitank:
      73mm, 82mm, 107mm
          Recoilless Rifles
      85mm, 100mm, 105mm guns
      Sagger, SS–11, Milan, HOT
          Missiles

Air Defense:
4000 23mm, 37mm, 57mm, 85mm,
100mm, 130mm Guns
SA–2/3/6/7/9 SAM
30 Roland

		Personnel
Navy		4,500
1	Frigate	
10	Osa FAC	
5	Large Patrol Craft	
12	P–6 FAC	
10	Coastal Patrol Craft	
5	Minesweepers	
4	LST	

Air Force		38,000
2	Bomber Squadrons	
	7 Tu-22	
	8 Tu-16	
11	Fighter Squadrons	
	100 MIG–23	
	95 Su–7	
	80 Su–20	
	12 Hunter	
	5 Super Etendard	
5	Interceptor Squadrons	
	25 MIG–25	
	40 MIG–19	
	150 MIG–21	
	49 Mirage	
1	Recce Squadron	
	5 Mig–25	
2	Transport Squadrons	
	10 An–2	
	10 An–12	
	8 An–24	
	2 An–26	
	9 Il–76	
	2 Tu–124	
	13 Il–14	
	1 Heron	
11	Helicopter Squadrons	
	35 Mi–4	
	15 Mi–6	
	150 Mi–8	
	40 Mi–24	
	40 Alouette II	
	11 Super Frelon	
	50 Gazelle	
	13 Puma	
	30 BO–105	
	7 Wessex Mk 52	

ISRAEL

	Personnel
Army	104,000

Organization
11 Armored Divisions
33 Amored Brigades
10 Mechanized Infantry Brigades
12 Territorial Infantry Brigades
15 Artillery Brigades

Equipment
Tanks:
1100 Centurion
600 M–48
1210 M–60
250 T–54/55
150 T–62
250 Merkava I/II
Armored Fighting Vehicles:
4000 Ramta RBY, BRDM–1/2
4000 M–113, OT–62, BTR–50 APC
Artillery:
140 175mm SP
70 105mm
100 122mm
300 Soltam L–33
300 155mm SP
48 203mm
— Lance SSM
— 122mm, 160mm, 240mm,
290mm MRL
Antitank:
— 82mm Rocket Launcher
— 106mm Recoilless Rifle
— TOW, Dragon, Picket, Milan
Missiles
Air Defense:
24 Vulcan/Chaparral
900 20mm, 23mm, 30mm, 40mm
Guns
— Redeye

Navy		9,000
3	Type 206 Submarines	
4	Aliya Corvettes	
20	Fast Attack Craft	
2	Flagstaff Hydrofoil	
45	Coastal Patrol Craft	
12	Amphibious Ships	

	Personnel
Air Force	28,000

19 Fighter Squadrons
   40 F–15
  131 F–4E
  150 Kfir
   67 F–16A
    8 F–16B
  130 A–4N/J
Recce:
  13 RF–4E
   2 OV–1E
   4 E–2C
   4 RU–21J
   2 C–130
   4 Boeing 707
1 Transport Wing
   7 Boeing 707
  20 C–130 E/H
  18 C–47
   2 KC–130H
Helicopters:
  30 AH–1S
  30 Hughes 500MD
   8 Super Frelon
  33 CH–53A
   2 S–65E
  25 Bell 206
  60 Bell 212
  17 UH–1D
15 Surface-to-air Battalions
    Improved Hawk

NORTH YEMEN

		Personnel
Army		35,000

Organization
1 Armored Brigade
1 Mechanized Brigade
5 Infantry Brigades
1 Special Forces Brigade
1 Para/Commando Brigade
1 Central Guard Force
3 Artillery Brigades
3 Antiaircraft Artillery Battalions
2 Air Defense Battalions
Equipment
Tanks:
  100 T–34
  500 T–54/55
   64 M–60
Armored Fighting Vehicles:
  50 Saladin, Ferret, BMP Cars

  90 M–113 APC
300 BTR–40/60/152 APC
Artillery:
200 76mm, 105mm, 122mm, 155mm Howitzers
  30 SU–100 SP
200 82mm and 120mm Mortars
  65 BM–21 Rocket Launcher
Antitank:
  — 75mm, 82mm Recoilless Rifles
  — LAW Rocket Launcher
  20 Vigilant Missiles
  24 Dragon
  — TOW
Air Defense:
  — ZU–23 23mm, 37mm, 57mm, 85mm Guns
   4 ZSU–23–4
  72 Vulcan 20mm
  — SA–2/7 SAM

Navy		550

4 P–4 FAC
6 Patrol Craft
2 T–4 Landing Craft

Air Force		1,000

5 Fighter Squadrons
  40 MIG–21
  10 MIG–17F
  11 F–5E
  15 Su–22
Transport:
   2 C–130H
   2 C–47
   1 Il–14
   1 An–24
   3 An–26
Helicopters:
  20 Mi–8
   6 AB–206
   5 AB–212
   2 Alouette
1 Air Defense Regiment
  12 SA–2 SAM

SOUTH YEMEN

		Personnel
Army		24,000

Organization
1 Armored Brigade
1 Mechanized Brigade

10	Infantry Regiments	
1	Artillery Brigade	
10	Artillery Battalions	
1	Surface-to-Surface Missile Brigade	

Equipment

Armored Fighting Vehicles:
- 450 T–54/55/62 Tanks
- 100 BMP Mech Inf Veh
- 300 BTR–40/60/152 APC

Artillery:
- 350 85mm, 100mm, 122mm, 130mm Guns
- — BM–21 122mm MRL
- 12 FROG–7
- 6 Scud

Air Defense:
- 200 ZU–23/ZSU–23 23mm, 37mm, 57mm
- 6 SA–2 SAM
- 3 SA–3 SAM
- — SA–6/7 SAM

Navy   1,000
- 1 Corvette
- 8 Osa FAC
- 2 Large Patrol Craft
- 2 P–6 FAC
- 2 Zhuk FAC
- 1 LST
- 3 LCT
- 3 LCA

Air Force   2,500
- 4 Fighter Squadrons
  - 30 MIG–17
  - 12 MIG–21
  - 25 Su–20/22
- 3 Interceptor Squadrons
  - 36 MIG–21F
- 1 Transport Squadron
  - 3 An–24
- 1 Helicopter Squadron
  - 15 Mi–25
  - 30 Mi–8
- 1 Surface-to-air Missile Regiment
  - 48 SA–2

AUTHORS

Brigadier General Ahmed M. Abdel-Halim is an Egyptian armored/operations officer. General Halim earned his BS from the Egyptian Military Academy and an MA from the American University in Cairo. In addition, he has attended the Egyptian Command and Staff College and held a fellowship at the Egyptian Higher War College. The general has been an instructor at both the Egyptian Armoured School and the Egyptian Armoured Institute and has served as an Egyptian Military Attaché to Somalia. Prior to attending the National Defense University, General Halim was the Deputy of Planning in the Operations Department of the Egyptian Ministry of Defense. In addition to this monograph, General Halim is the author of the forthcoming *Iraq-Iran War (3 Levels: Strategic, Operational and Tactical).*

Brigadier General Yehuda Bar is an infantry commander with the Israeli Army. In addition to attending the National Defense University, General Bar has attended the Command and Staff College in Israel, the United States Marine Corps Command and Staff College, and Tel Aviv University. Prior to his participation in the International Fellows Program General Bar was in command of an infantry division in both Lebanon and Israel.

Colonel Lee Suk Bok is a South Korean officer specializing in Field Artillery and Operations. Colonel Lee received a BS from the Korean Military Academy and is a graduate of the Army Staff College in Korea. He has attended several basic and advanced officer courses in field artillery in both the United States and South Korea. Colonel Lee has had teaching experience as the Chief Instructor at the Air Defense School in Korea and was the division artillery commander of the 20th Mechanized Infantry Division before accepting an invitation to attend the National Defense University.

Brigadier Zia Ullah Khan is an infantry commander with the Pakistani Army. General Zia, a graduate of Emerson College in Lahore, Pakistan, was the commander of the 20th Infantry Brigade in Pakistan before attending the International Fellows Program. Prior to assuming command of the 20th Brigade, Brigadier Zia served as a divisional staff officer and was an instructor at the School of Tactics and Infantry in Pakistan.

Colonel Abdulaziz bin Khalid Alsudairy is a member of the Royal Saudi Air Force. A graduate from the British Royal Air Force Academy in Cranwell and the British Royal Air Force Staff College in Bracknell, Colonel Alsudairy was the Commanding Officer of the Flying Wing at King Fahad Air Base in Taif, Saudi Arabia, before attending the National Defense University. In addition, Colonel Alsudairy has served as both Operations and Commanding Officer for the Second Squadron at King Faisal Air Base in Tabuk, Saudi Arabia.